EVERY WAR MUST END

Columbia Classics

Every War Must End

Second Revised Edition
With a New Preface

FRED CHARLES IKLÉ

COLUMBIA UNIVERSITY PRESS
NEW YORK

Columbia University Press
Publishers Since 1893
New York   Chichester, West Sussex
Copyright © 1971, 1991, 2005 Columbia University Press
All rights reserved

Library of Congress Cataloging-in-Publication Data

A complete CIP record is available from the
Library of Congress.
ISBN 978–0–231–13666–2 (cloth)
ISBN 978–0–231–13667–9 (paper)

♾

Columbia University Press books are printed
on permanent and durable acid-free paper.

Printed in the United States of America

CONTENTS

Preface to the Second Revised Edition   vii

Preface to the Revised Edition   xvii

One   The Purpose of Fighting   1

Two   The Fog of Military Estimates   17

Three   Peace Through Escalation?   38

Four   The Struggle Within: Patriots Against "Traitors"   59

Five   The Struggle Within: Search for an Exit   84

Epilogue   Ending Wars Before They Start   106

Notes   133

Bibliography on the Termination of Wars   151

Index   155

PREFACE TO THE
SECOND REVISED EDITION

History is a cruel tutor. It hammers a lesson into our minds so sternly that no one dares to mention the many exceptions that must be allowed. Yet as soon as we have learned that lesson—and ignored its exceptions—history punishes us for not following another rule that posits the very opposite.

Since the Second World War, the lesson that an aggressor must not be appeased has been evoked again and again. Anyone who argued that an aggressor might be kept at bay with some limited concessions risked being called a Neville Chamberlain, the prototype appeaser traveling with his umbrella to Munich to give in to Hitler. Then, in the second half of the twentieth century, history tutored a different rule: Do not start a war unless you are sure of winning it. America's protracted and painful struggle to end the war in Vietnam taught us that lesson—with blood, tears, and political turmoil at home. Thereafter, all the backward-looking pundits stopped accusing political leaders of "another Munich!" and switched to warning of "another Vietnam!" Some commentators even called the Soviet Union's war in Afghanistan "a Soviet Vietnam," despite the many moral and political differences of that war.

President Lyndon Johnson found himself torn by history's contradictory lessons. He had inherited the Vietnam war, but his attempt to end it by increasing America's military commitment failed. He was thus forced to conclude that he could not, and should not, seek reelection. After his retirement he told his biographer, Doris Kearns: "Oh, I could see it coming all right. History provided too many cases where the sound of the bugle put an immediate end to the hopes and dreams of the best reformers: the Spanish-American war drowned the populist spirit; World War I ended Woodrow Wilson's New Freedom. . . . Yet everything I knew about history told me that if I got out of Vietnam and let Ho Chi Minh run through the streets of Saigon, then I'd be doing exactly what Chamberlain did in World War II."[1]

America's war in Vietnam ended with the fall of Saigon in 1975. The memory of that agonizing search for an exit weighed on U.S. policymakers for decades, like a recurring nightmare. It might have contributed to the U.S. decision to end the Gulf War as soon as the Iraqi invaders had been driven out of Kuwait. That benchmark met the stated war aims of the United States and the United Nations. Colin Powell, who had served twice in Vietnam, was Chairman of the Joint Chiefs of Staff during the Gulf War. After Iraq's invasion of Kuwait, in August 1990, he advocated—and guided—a huge military buildup, so that the United Sates would have preponderant military strength in the Persian Gulf region should war with Iraq become necessary. Thanks to his prudent planning, the United Sates and its allies were fully prepared to wage the amazingly successful, short campaign early in 1991 that expelled Saddam's best forces from Kuwait.

In his autobiography Colin Powell recalls:

Before the war began, someone on my staff had given me a book entitled *Every War Must End*, by Fred Iklé. I had worked with Iklé when he was undersecretary of defense for policy and I was Cap Weinberger's military assistant. The theme of his book intrigued me, because I had spent two tours in a war that seemed endless and often pointless. Warfare is such an all-absorbing enterprise, Iklé wrote, that after starting one, a government may lose sight of ending it. . . . I was so impressed by Iklé's idea that I had key passages photocopied and circulated to the

---

1. Doris Kearns, *Lyndon Johnson and the American Dream* (New York: Harper & Row, 1976), 252.

Joint Chiefs, Cheney and Scowcroft. We were fighting a limited war under a limited mandate for a limited purpose, which was soon going to be achieved. I thought that the people responsible ought to start thinking about how it would end.[2]

On February 27, 1991, as the last of the defeated Iraqi forces fled from Kuwait into Iraq as fast as they could, Powell was called to a meeting with the president and his senior advisors to discuss how to end the war. On the way to this meeting, Powell writes, "words from Iklé's book ran through my mind: '. . . fighting often continues long past the point where a rational calculation would indicate the war should be ended.'" Powell wrapped up his briefing to the president by saying that tomorrow he would bring him a recommendation to stop the fighting. "If that's the case," the president responded, "why not end it today?"[3]

With the benefit of hindsight, critics have since argued the United States should not have ended the Gulf war so suddenly but should instead have used its victorious forces to destroy more of Saddam Hussein's military assets, or even to march on to Baghdad. Although most of America's allies would not have supported such an expansion of the war, it seems plausible that the United States could have occupied all of Iraq quickly and thus ended Saddam Hussein's brutal rule in 1991. Then what?

The Shiites would probably have rallied to achieve a leading role in the new Iraq, and the Sunni followers of Saddam—angry and driven out of office—might have begun to oppose the American occupation. Thus, had the United States continued its military offensive in 1991 and succeeded in removing Saddam Hussein, the aftermath of expanding the Gulf War might not have been an enduring peace in Iraq but a protracted insurgency, perhaps similar to the insurgency that confronted America's forces after the invasion of 2003. This is a conjecture, of course, the kind of speculation known as "counterfactual history."

When statesmen face choices about war and peace, they cannot dwell at length on the wrong turns they, or others, might have taken in the

---

2. Colin L. Powell, *My American Journey* (New York: Random House, 1995), 519.
3. Powell, *My American Journey*, 521.

past. They must look ahead. Nevertheless, they will be strongly influenced by lessons of history. As Lyndon Johnson recalled, his thinking was influenced by Woodrow Wilson's and Neville Chamberlain's mistakes; Colin Powell indicated that his approach was informed by America's mistakes and travail in Vietnam. If we rely on lessons of history, we ought to keep in mind that the bottom line of these lessons is always a comparison of two parts. One part is the actual history: the known facts of what did happen following the decision that was taken. The other is counterfactual history: conjectures about what might have happened along the road not taken. This mixture of facts and speculation is unavoidable. It is embedded in any broad assessment of policies and strategies that have been pursued in past wars. It surely affects what can be said about America's second war against Saddam Hussein.

For twelve years after the end of the Gulf war, Saddam Hussein violated not only most arms-control provisions of the cease-fire agreement with the United States but also ignored seventeen United Nations resolutions ordering Iraq to comply. Moreover, his continuing violations suggested to American intelligence specialists that he was hiding a program to develop weapons of mass destruction. And because of the psychological impact of the 9/11 attack, Americans saw this intelligence about Iraq (which turned out to be false) in the darkest colors. For victims of a recent tornado, a rapidly darkening sky is a warning to be heeded promptly. America's patience with the Iraqi dictator had run out.

When should a nation at peace start a war to remove what it perceives as an imminent threat? When should a nation at war seek peace to achieve the best available outcome? Such choices demand the most consequential, the most agonizing decisions that leaders of a democracy might face. The memory of these epochal choices resonates for a long time—a haunting reminder that the most fateful decisions are irreversible. Robert Frost's poem about "the road not taken" speaks to this:

> Two roads diverged in a yellow wood,
> And sorry I could not travel both.
> . . . .
>
> Yet knowing how way leads on to way,
> I doubted if I could ever come back.

> I shall be telling this with a sigh
> Somewhere ages and ages hence:
> Two roads diverged in a wood, and I—
> I took the one less traveled by,
> And that has made all the difference.

What "has made all the difference" for the long-term outcome of many wars is whether the militarily victorious side managed to reform the enemy's government, to transform a former foe into a new friend.

On this point, the contrast between the peace settlements of World War I and those of World War II is stunning. The victors of the First World War sought a punitive ending. France, Great Britain, and (more reluctantly) the United States imposed heavy reparations on defeated Germany, sliced off parts of German territory, and presided over the Balkanization of the Hapsburg empire. But they paid no attention to the impact of the reparations on the evolution of Germany's postwar government and gave little thought to new ethnic clashes that their arbitrary redrawing of national boundaries might provoke. John Maynard Keynes predicted, with enviable prescience, the catastrophic consequences of this punitive peacemaking. Already in 1919, he wrote that the peace treaty imposed on Germany "includes no provisions for the economic rehabilitation of Europe,—nothing to make the defeated Central Empires into good neighbors, nothing to stabilize the new States of Europe, nothing to reclaim Russia."[4]

Having sown the wind of vengefulness, the victors of 1918 reaped the whirlwind twenty years later. But by that time, the United States and England had learned the lesson. Starting with the Atlantic Charter in August 1941, Franklin Roosevelt and Winston Churchill laid down principles for a more prudent way to end the Second World War. The reward for the strategic foresight of these statesmen was, above all, the creation of stable democracies in Germany and Japan. In addition, in 1944 the United States, Great Britain, and their allies agreed on new economic institutions (such as the World Bank) that helped rebuild the global econ-

---

4. John Maynard Keynes, *The Economic Consequences of the Peace* (New York: Harcourt, Brace & Howe, 1920), 226.

omy and undergirded the Marshall Plan. These successes, of course, could not penetrate the Iron Curtain and thus did not avert the Cold War.

In 2003, when American and British forces fought the second war against Saddam Hussein, they defeated him militarily at a much smaller cost in casualties and destruction than they had feared. Alas, as soon as the mobs in Baghdad had toppled the statues of their fallen dictator, a violent, prolonged, and painful struggle for winning the peace began to darken that splendid military victory. To be sure, officials within the Bush administration had prepared thoughtful plans for supporting Iraq's economic reconstruction, phasing out the military occupation, and guiding the political development of a new Iraq. So what went wrong?

At the time of this writing it is too early to give a full and fair answer. But from the assessments and data available so far, we can perhaps derive a few tentative rules, applicable to similar situations in the future. One rule seems elementary: Democracies that have achieved a military victory ought to refrain from seeking revenge. Taking revenge is a Neanderthal strategy. Instead of giving priority to a policy that can transform the defeated enemy into an ally, the revenger helps the hawks on the enemy's side to recruit angry fighters who will undermine the peace settlement. During the critical weeks following the collapse of Saddam Hussein's rule in Baghdad, the emphasis on punishment and revenge clearly harmed America's long-term objectives. For instance, no attempt was made to arrange a formal surrender with one of the senior Iraqi officials who turned themselves in voluntarily, Saddam Hussein's foreign minister, Tariq Aziz. The U.S. decision makers in Baghdad and Washington deliberately inflicted indignities on nearly all senior Iraqi generals and Ba'athist officials and had them imprisoned for years under humiliating conditions. To what end?

The Unites States should have tried promptly to recruit officers from Saddam Hussein's collapsed command structure to form a reformist Ba'ath cadre and use reeducation, iron discipline, and attractive inducements to split off an ever larger number of the Sunni leadership from the die-hard opponents of a new democratic Iraq. Obsessed with a desire to punish and revenge, the U.S. managers of Iraq's occupation delayed this approach for more than a year, at which time the United States was confronted by an organized, hostile insurgency.

The psychological impact among Americans of the 9/11 attack might have contributed to the mistaken vengefulness following America's military victory in Iraq. The injury and insult of 9/11 justifiably enraged people throughout the United States, a rage that senior officials tried to address by promising that the terrorists "will be brought to justice." To be sure, "bringing to justice" can give satisfaction to victims who find year-long jury trials healing. But the struggle against global terrorism will not be won by court trials.

The successful ending of World War II owes much to the Allies' prudence and skill in promoting formal acts of surrender—in Italy, even in Germany, and above all in Japan. These surrenders were carried out between senior military commanders as well as between representatives of the central governments. The lesson of Japan's surrender in 1945 is particularly instructive. The hawks on both sides almost succeeded in prolonging the war (see pp. 71 and 93 below). For the Allies, that would have meant huge additional casualties in protracted guerilla fighting throughout the many islands and territories still occupied by fanatically loyal Japanese soldiers. Moreover, such a continued distraction of the United Sates and Great Britain would have been gleefully exploited by Stalin to make further territorial gains. At the same time, Japanese society would have suffered huge additional casualties and a further loss of its cultural heritage. The resultant despair and hatred in Japan would have delayed its emergence as a wealthy, stable democracy or might even have derailed Japan's democratization indefinitely.

Another principle essential for turning a military victory into a lasting political success has to do with the prestige that the victor's forces ought to gain—and maintain—among the defeated population. The morning after all of Saddam's statues had tumbled into the dust, the American forces seemed ten feet tall to many Iraqis. From what the Iraqi people had seen, they came to believe the American military could find any target, day or night, and hit it; could rapidly repair the electric system and water supply (if they wanted); and would be the new saving force to maintain law and order effectively and justly. A day or two later, when mobs of looters ran through the streets of Baghdad, these proud, all-powerful American forces looked on sheepishly and did nothing to kill this incipient uprising in the bud. The looters not only emptied museums and shops, but they stole office supplies from government buildings that were needed for restoring a functioning administration and even gathered up intelligence documents that would have been invaluable for tracking down or intimidating future troublemakers. To boot, as these scenes of chaos were broadcast throughout the Islamic world, some senior officials in Washington chuckled about a "new spirit of freedom" that had suddenly sprouted (in the disguise of looting) among "grateful," liberated Iraqis. America lost most of its prestige and respect in that episode. To pacify a conquered country, the victor's prestige and dignity is absolutely critical. General Douglas MacArthur knew this.

It remains to be seen what future assessments of the war and the subsequent insurgency in Iraq might teach us about winning a military campaign without losing the peace. For some time to come, America's policymakers will be more hesitant to initiate a major military action, as they will be reminded of the pain and suffering of the brave U.S. military forces who strove to pacify Iraq. But the world is getting too dangerous for defeatism. The proliferation of weapons of mass destruction is accelerating and growing into an elemental threat to all nations. A dictatorship that acquired these weapons might sell them to terrorist organizations or start using them to unhinge the international order. United Nations resolutions would be ineffective in averting such a calamity, unless they were promptly enforced by effective military action. This much, Saddam Hussein has taught us.

But apart from the United States, few nations will have the military capability to act as enforcers. And many nations will hesitate to step forward so long as they fear getting drawn into "another Iraq." At that point, waiting for the United Nations to assemble a peacekeeping force would push the world over the brink of disaster. And it would be escapist to repeat the hackneyed call for "another Marshall Plan" or to appeal to the ethereal "international community" to come to the rescue. In the turbulent era ahead, crises will occur that threaten national survival, and utter destruction might not be averted without the prompt use of military force. It is crucial, therefore, that the United States and its friends relearn the rules for ending a war with strategic foresight and skill so that the hard-won military victory will purchase a lasting political success.

PREFACE TO THE REVISED EDITION

Wars tranform the future. They move boundaries, topple governments, expand or break up empires, and leave scars of death and destruction. The battles fought during a war, of course, contribute to its aftermath; but it is the way in which a war is brought to an end that has the most decisive long-term impact. Yet, historians, foreign affairs experts, and military strategists have devoted far more thought to the question of how and why wars begin. While the origins of World War I and World War II, for example, have been studied in painstaking detail, the planning and efforts for terminating these wars have received much less attention.

This imbalance in the understanding of past wars affects how political leaders and military planners will approach questions of war and peace in the future. Regarding the beginning of wars, they can call on historic data, rich concepts, and extensive prior planning: how to deter aggression, how diplomacy might avert the outbreak of war, how to mobilize forces, and how to design the initial military campaigns. Much less is known about how to bring a war, once started, to a satisfactory end.

I worked on this book when American involvement in the war in Vietnam had become an agonizing search for an exit. The United States had become entrapped in painfully prolonged fighting, unable to marshal a strategy to end the war satisfactorily. But in the chapters written at that time, I never mention the war in Vietnam. If one seeks to draw lessons from history about questions of war and peace, references to current events can easily take on a disproportionate weight. The influence of the present is strong enough without quoting from today's newspapers.

I began to write this new introduction as the fighting in the six-week war for the liberation of Kuwait from Iraqi forces came to an end. Sixteen years have elapsed between this war in the Persian Gulf and the fall of Saigon in 1975 that ended America's war in Vietnam. During this time, about a dozen wars have broken out and some are still continuing after years of fighting. Indeed, this period disclosed new lessons about how much easier it is to begin a war than to end it.

Judged a bit coarsely, the United States lost the Vietnam War. The Gulf War was clearly won by the United States—if judged in military terms. In political terms, though, the verdict must wait. In the Vietnam War, the victory won by America's enemy was a Pyrrhic one. Hanoi, soon after its victory, became embroiled in prolonged hostilities in neighboring Cambodia from which it is still trying to extricate itself. Then, in 1979, Hanoi had to fight off a military incursion from China—its staunch ally in the long war against the United States. And all this time, Vietnam's economy deteriorated so much that "victorious" Hanoi is now humiliatingly dependent on foreign aid, over which Washington—the "defeated" enemy—has much to say. Could it now be said that by seeking South Vietnam's total defeat and the humiliation of the United States, Hanoi, in fact, lost the peace?

The Soviet Union, too, has recently experienced great difficulties in trying to end the involvement of its forces in a protracted war. Seven years after American forces had been withdrawn from Vietnam, Soviet forces invaded Afghanistan, only to become enmeshed in fighting that continued inconclusively for over eight years. It would be frivolous, however, to construct all sorts of analogies between the war in Vietnam and the Soviet invasion and fighting in Afghanistan. The origins of these two wars are vastly different, and so is the way in which each of the superpowers ended the involvement of its combat forces. The

Soviet-supported government in Kabul still controls the Afghan capital plus considerable territory, and continues to receive massive assistance from the Soviet Union. By contrast, after the withdrawal of American forces from Vietnam in accordance with the Paris peace accords of 1973, the United States Congress severely curtailed assistance to Saigon for munitions and fuel, without which the South Vietnamese army could no longer defend itself against the North Vietnamese forces who were invading the South (in clear violation of the peace agreement).

Moscow's difficulties in ending the fighting by Soviet forces in Afghanistan surely contributed to the far-reaching changes in Soviet foreign policy after 1988. Evaluation of the role played by Moscow's forces in Afghanistan continues among Soviet officials, military experts, and historians. But one conclusion appears to be widely accepted in Moscow among reactionary hard-liners as well as among democratic reformers. Brezhnev's decision to send forces into Afghanistan was thoughtlessly taken, with no plan for ending the adventure and little attention to the political goals of the invaison. As Soviet Foreign Minister Shevardnadze put it: "The people who made the decision about intervening with armed force did not plan to stay in Afghanistan for any length of time, or to create the 16th or 17th Soviet republic. . . . [it was] an incorrect, and badly thought out decision."[1]

America's travail in ending the war in Vietnam has profoundly influenced American thinking about questions of war and peace. Even though the historical assessment of the Vietnam War is still being debated in the United States among politicians, soldiers, and scholars, important lessons for American strategy have clearly been learned.

According to a simplistic interpretation of the Vietnam experience, the United States should have used military force more massively early on during its military involvement in Vietnam. Also, the American military leadership should have been given freer rein by the President to pursue the war against the enemy without granting him sanctuaries and without restricting what could be attacked. The war could then, it is argued, have been ended quickly and favorably. But this is not the correct lesson of Vietnam. On the contrary, as the following chapters seek to show, restraint applied to both the means and ends of warfare is essential to reaching a successful outcome in most wars. America's conduct of the war in Vietnam suffered not so much from limits imposed on the use of military force as from lack of an overarching strategy for applying military force in a way that would bring the war to a satisfactory end. As Richard Nixon put it: "Never in history has so much power been used so ineffectively as in the war in Vietnam."[2]

In essence, the first lesson of Vietnam is that American forces must not be committed to combat without a clear military strategy, whether for defeating the enemy or for expelling the aggressor's forces and restoring the peace.

A second and corollary lesson is that American forces should not be sent into combat merely for the purpose of demonstrating America's resolve and commitment. Such a "demonstration strategy" is no substitute for a clear military strategy designed to defeat the enemy's forces. It will not induce a determined adversary to withdraw or to cease his aggression.

However, despite the lesson of Vietnam, policymakers continue to be tempted by ideas for committing military forces to demonstrate resolve in an armed conflict where they cannot marshal the military strength—or the resolve—to win. In 1983, the terrorist attack in Beirut that killed 241 American servicemen provided a sad reminder of the futility and dangers of such a "demonstration strategy." Defense Secretary Caspar Weinberger had strongly counseled against the deployment of American forces that led to this tragedy, forces that were sent into Lebanon without a military strategy and without an achievable mission. Unfortunately, his counsel did not prevail until after the Beirut disaster when the American forces were at last withdrawn.[3]

A third lesson tells us that the United States should not enter a war based on a strategy of inflicting "punishment" on the enemy by bombing on shelling targets whose destruction will not serve to defeat the enemy's forces militarily. This lesson, in fact, should have been learned long before the Vietnam War, from the many thoughtful and detailed studies of "strategic" bombing in World War II that were undertaken in the late 1940s and early 1950s. To bomb some cantonments here, some military industries there, to demolish bridges here or buildings there, will not help to end a war unless the damage inflicted directly supports military campaigns that are designed to defeat the enemy's forces. A despotic ruler will not sue for peace merely because his soldiers or his civilians suffer pain and death. The "punishment strategy" could not end the Korean War against Kim Il Sung or the Vietnam War against Ho Chi Minh; and had the United States relied on it in the Gulf War against Saddam Hussein it would have failed there, too.

Inflicting "punishment" on the enemy nation is not only an ineffective strategy for ending a war, it may well have side effects that actually hasten the defeat of the side that relies on such a strategy. Two and a half years into World War I, Germany decided it needed a new strategy to bring the war to a satisfactory end quickly. It launched an unlimited submarine campaign against England with the idea that this would "cause panic" among the English people. Instead, the strategy brought the United States into the war against Germany and thus led to Germany's defeat (see pp. 42–49).

When, on the other hand, attacking targets does serve a sound military strategy—rather than a vague notion of punishment—it is important for the United States and other modern democracies to avoid inflicting civilian casualties and needless damage. Unless governments of modern democracies can demonstrate that they are expending every effort to avoid unnecessary destruction, they will lose the necessary political support for pursuing the war to a satisfactory end, at home as well as among their allies. In the last analysis, democracies must avoid wanton damage not only to maintain public support for the war effort but also to conduct the war in a way that is consonant with the nation's basic values.[4]

In the six-week war in the Persian Gulf, American strategy sought to heed every one of these lessons. The U.S.-led coalition assembled sufficient force to expel the Iraqi army from Kuwait, and from beginning to end it followed a comprehensive and careful war plan to achieve this military objective. Moreover, the tactics and weaponry of the American forces and the coalition partners destroyed military targets with precision—targets selected so that their destruction would make a direct contribution to the military campaign while carefully avoiding, as much as possible, civilian damage.

Interestingly, the five-month period of United Nations sanctions against Saddam Hussein's Iraq in a sense tested the "punishment" strategy. Unlike the "punishment" inflicted by indiscriminate bombing, these economic sanctions, of course, did not physically destroy Iraq's homes and civilian industries but were designed to cripple its economy and prevent further arms imports. Many members of Congress wanted to postpone the recourse to force in the hope that a continuation of these sanctions would eventually compel Saddam Hussein to withdraw from Kuwait. After an intense and thoughtful debate, however, a slight majority in both Houses supported President Bush's request to endorse the use of force to expel the Iraqi army.

In hindsight after the six-week war, it now seems highly implausible that the economic punishment strategy would have worked. Since Saddam Hussein stubbornly refused to yield on Kuwait as he was rapidly losing his military strength, week after week, it is hard to imagine that he would have yielded merely because the sanctions made it difficult for him to import additional weaponry.

A delicate balance must be struck in trying to cut off outside military assistance to the enemy. On the one hand, continued and substantial outside support (especially if furnished by major powers with seemingly unlimited resources) will strengthen the hawkish sentiment among the enemy leadership, overcoming more cautious judgments in favor of a negotiated settlement. During the later part of the Vietnam War, the generous flow of military assistance from the Soviet Union and China to the North Vietnamese forces undoubtedly contributed to Hanoi's recalcitrance in the peace negotiations with Washington. On the other hand, an attempt to interdict the outside flow of aid to an enemy's forces might widen the war instead of bringing it to an end. It can provoke a nation that has been sending aid to the enemy into joining the war on the enemy's side.

A recent example of a blockade that came close to triggering such an expansion of a war is furnished by the Iran-Iraq War in the 1980s. Iran attempted to tighten the naval blockade against Iraq by attacking Kuwaiti and other third party shipping in the Persian Gulf. This led to the American agreement with Kuwait in 1987 to convoy oil tankers. The United States then obtained support from allied naval units, strengthened its military cooperation with the Gulf states, established necessary coordination with the Iraqi military, and engaged in several skirmishes with Iranian forces that seriously weakened the small Iranian navy. Fortunately, Iranian leadership sensed the danger Iran was courting in trying to end the war with Iraq by expanding the naval blockade in the Gulf: in August 1988, it agreed to a United Nations sponsored cease-fire with Iraq. The speaker of the Iranian parliament, Hashemi Rafsanjani, said on July 29, 1988: "Our acceptance of [the U.N. cease fire] must not be interpreted as if it were a compromise or submission to extortion. This new move puts an end to our unnecessarily making enemies." And the Iranian deputy foreign minister, Larijani, explained the acceptance of the cease fire by pointing out that Iran's war with Iraq at first was "a bilateral conflict unleashed by Iraqi aggression; then it became regional; until one day we noticed that it had now taken on an international dimension."

During the same years, Moscow decided to end the war that its forces were fighting in Afghanistan. Here, too, an attempt might have been made by the Soviet Union to cut off the outside aid the Afghan resistance was receiving. The steadily increasing arms deliveries to the resistance that the United States and other countries arranged were imposing ever higher costs on the Soviet military.[5] Hence, the Soviet military might well have been tempted to expand the war in order to interdict this arms flow, in particular by attacking supply lines in Pakistan.[6] Happily, instead of widening the war beyond Afghanistan and risking a vastly expanded conflict, Moscow decided to end its direct role in the fighting by withdrawing its forces.

Outside support is particularly difficult to cut off for guerrilla forces. The most protracted wars in recent decades have been precisely those fought as a kind of civil war where insurgent forcces, who often blend into the civilian population, use hit-and-run raids and terrorist acts to undermine the authority of the central government. The difficulty of ending insurgency wars can be seen today in many countries. Colombia has suffered from terrorist insurgencies for many decades. On a smaller scale and more sporadically, the people in Northern Ireland have long been the victims of such warfare by the IRA. And for ten years now the government of El Salvador has fought a most tenacious insurgent movement.

Not only the central government combating a guerrilla insurgency may find it difficult to bring such a war to an end. If a guerrilla movement succeeds in gaining increasing control of the nation's territory it may itself become vulnerable to the hit-and-run attacks, sabotage, and terrorism that it has practiced so successfully. The tragic ebb and flow of this kind of warfare can thus lead to ever more brutal terrorism by the guerrilla forces and ever more cruel repression by the forces seeking to establish territorial control. During the 1960s and 1970s, a pessimistic view gained ground in Western democracies about this contest between insurgent movements and the efforts of governments to end the fighting and maintain territorial control. The side that resorted to totalitarian rule, systematic terror, and mass murder would always win, while democratic governments trying to control totalitarian insurgencies or democratic movements struggling to overthrow tyrannical dictatorships would always lose. For example, the British government has so far found no solution for ending IRA terrorism that would be compatible with its political and legal values. By contrast, to suppress a Kurdish insurgency in 1988, Saddam Hussein could use poison gas and mass terror with impunity.

Since the end of the Cold War, however, a sense of optimism has welled up among the nations governed by democratic principle. In the short run, tyrants may prevail; but in every corner of the globe vast majorities are winning the slow struggle to establish democratic governance. In the view of many, the global spread of democracy is creating an expanding realm of peace: modern democracies will not and cannot wage war against each other.

Yet this global spread of democracy will not be without reversals. We must expect that nations, ethnic groups, religious or political movements will again come under the control of tyrants who unhesitatingly start wars to expand their dominion or to destroy their adversaries. At the same time, we can see no escape from our age of technology, an age that provides such tyrants with access to weapons of mass destruction—nuclear, chemical, biological—and with ever more varied ways of employing them. How could democracies end a war forced upon them by such a tyrant? He might have reached a state of mind so that, unable to find an exit from the war he started, he decides to use his dreadful arsenal—not to preserve his rule, but only to slaughter his enemies; not to expand his sway or to achieve any kind of victory, but only for revenge. As if to warn us of things to come, some of Saddam Hussein's threats in the Gulf War could have been motivated by such a state of mind.

In a future war, to confront an aggressor so motivated and armed with mass destruction weapons, surely, we could not rely on "graduated deterrence," "flexible response," and similar strategic concepts from the Cold War era.[7] We need a new strategy to supersede the concepts we developed for the bipolar confrontation between Western democracies and "the Communist Bloc." Any future tyrant who would launch a war of aggression, regardless of cost or consequences, will have to be deprived of mass destruction weapons before he uses them. To this end, the leading democracies will have to develop and acquire new military capabilities, and—more importantly—seek to create a new political order of the world. It will be the purpose of this endeavor to bring every war to an end without unleashing the cataclysmic destruction made possible by modern technology.

EVERY WAR MUST END

ONE

THE PURPOSE OF FIGHTING

*We accepted this war for an object—a worthy object—
and the war will end when that object is attained.
Under God I hope it will never end until that time.*
—ABRAHAM LINCOLN

*War's objective is victory—not prolonged indecision. In
war, there is no substitute for victory.*
—GENERAL DOUGLAS MACARTHUR

On s'engage et puis on voit.—NAPOLEON

FIGHTING A WAR can cost more in blood and money
than any other undertaking in which nations engage. And to wage
war, governments develop more detailed plans, establish a more
rigid organization, and institute tighter discipline than for any
other national effort. Yet, despite all this elaborate and intense
dedication, the grand design is often woefully incomplete. Usually,
in fact, it is not grand enough: most of the exertion is devoted to
the means—perfecting the military instruments and deciding on
their use in battles and campaigns—and far too little is left for relating these means to their ends.

In part, this deficiency stems from the intellectual difficulty
of connecting military plans with their ultimate purpose. Battles

and campaigns are amenable to analysis as rather self-contained contests of military power and, to some extent, are predictable on the strength of rigorous calculation. By contrast, the final outcome of wars depends on a much wider range of factors, many of them highly elusive—such as the war's impact on domestic politics or the degree to which outside powers will intervene.

In part, governments tend to lose sight of the ending of wars and the nation's interests that lie beyond it, precisely because fighting a war is an effort of such vast magnitude. Thus it can happen that military men, while skillfully planning their intricate operations and coordinating complicated maneuvers, remain curiously blind in failing to perceive that it is the outcome of the war, not the outcome of the campaigns within it, that determines how well their plans serve the nation's interests. At the same time, the senior statesmen may hesitate to insist that these beautifully planned campaigns be linked to some clear ideas for ending the war, while expending their authority and energy to oversee some tactical details of the fighting. If generals act like constables and senior statesmen act like adjutants, who will be left to guard the nation's interests?

WAR PLANS: PLANS WITHOUT AN ENDING

Three months before the attack on Pearl Harbor, the Emperor of Japan asked the Army Chief of Staff, Sugiyama, how long it would take the army to finish the job in the event of war with the United States. Sugiyama answered that operations in the South Pacific would be concluded in three months. The Emperor objected that when the war with China broke out Sugiyama had told him it would end in a month, yet after four years the fighting was still going on. Sugiyama's excuse was that the interior of China was huge; the Emperor replied in anger: "If the interior of China is huge, isn't the Pacific Ocean even bigger? How can you be sure that the war will end in three months?" [1]

By offering the excuse about the vastness of China, General

Sugiyama surely could not have meant that the Japanese military were unaware of the size of China when they started war with that country. The Japanese military leaders also knew the size of the Pacific before they attacked Pearl Harbor and they were, of course, fully aware of the industrial might of the United States. Since Japan became involved in a war with the United States neither gradually nor inadvertently, but by a considered and clear-cut decision, one would expect the Japanese military to have had some ideas about how they would reach a successful conclusion in the gigantic undertaking that they proposed.

On September 6, 1941, after the above exchange between the Emperor and the Army Chief of Staff, the proposal for attacking the United States was discussed further in a conference among the top military and civilian leaders. The Navy Chief of Staff recognized that Japan would have to be prepared for a long war. "Even if our Empire should win a decisive naval victory," he said, "we will not thereby be able to bring the war to a conclusion. We can anticipate that America will attempt to prolong the war, utilizing her impregnable position, her superior industrial power, and her abundant resources. Our Empire does not have the means to take the offensive, overcome the enemy, and make them give up their will to fight." The Navy Chief of Staff added that Japan would establish the basis for conducting a prolonged war by seizing strategic areas and resources at the outset. But "what happens thereafter," he went on, "will depend to a great extent on overall national power—including various elements, tangible and intangible—and on developments in the world situation." What an incredibly murky prospect for such a deep plunge!

It is not that the Japanese military had forgotten that the war they proposed to start must have an ending. The question was there, merely the answer was missing. A memorandum they had prepared for the conference of September 6 contained a long list of questions and proposed answers regarding the outlook for the German-Soviet war, the defense of the homeland, tactics for the ongoing negotiations in Washington, the strength of the United States army, and so forth. And one question in this list did raise the most crucial point: "What is the outlook in a war with Great Britain and the United States; particularly, how shall we end the war?" The Japanese military answered their own question as follows [italics added]: "A war with Great Britain and the United States will be long. . . . It is very difficult to predict the termination of a war, and it would be well-nigh impossible to expect the surrender of the United States. However, we *cannot exclude the possibility* that the war may end because of a great change in American public opinion. . . . At any rate, we should be able to establish an invincible position. . . . Meanwhile, *we may hope* that we will be able to *influence the trend of affairs* and bring the war to an end." [2]

To their credit, several Japanese leaders remained opposed to the Pearl Harbor attack. They did not dare, however, to press the military for the script of the last act. Instead of persistently asking how a war with the United States would end, they argued that the first act—the surprise attack—should be postponed a little while.

It is not uncommon among government officials (and probably among other administrators, as well) to turn the discussion of a controversial decision into a debate about its timing, while its merits are never dealt with. Japan's leaders, in considering the question of going to war against the United States, provide a stark example of this bureaucratic procedure.

In the conference of November 1, 1941, where the final decision in favor of war was taken, Army Vice Chief of Staff Tsukada argued: "The first things to decide are the central issues: 'to decide immediately to open hostilities,' and 'war will begin on the first of December.' . . . I would like to see diplomacy studied after these have been decided." At first, Foreign Minister Togo resisted any effort to settle the question of war and peace in the guise of settling the timing of an offensive: "You say there must be a deadline for diplomacy. As Foreign Minister, I cannot engage in diplomacy unless there is a prospect that it will be successful. I cannot accept deadlines or conditions if they make it unlikely that diplomacy will succeed. You must obviously give up the idea of going to war."

However, at this point a heated debate about timing ensued, and soon Foreign Minister Togo permitted himself to be diverted from the central issue. "November 13 is outrageous," Togo exclaimed (quibbling about a week), "the Navy says November 20." After further arguments about these dates, the Army Vice Chief of Staff conceded that November 30 "would be all right." Now, Foreign Minister Togo pleaded: "Can't we make it December 1? Can't you allow diplomatic negotiations to go on even for one day more?" Tsukada, the decision for war thus won by default, remained adamant: "Absolutely not, we absolutely can't go beyond November 30. Absolutely not." [3]

At last, they all agreed to delay the final step for setting the Pearl Harbor attack in train until midnight of November 30. This decision settled the momentous issue of whether or not Japan should attack the United States and neatly avoided the question of how the adventure would end.

The Pearl Harbor attack was one of the most successfully planned military operations in history. Yet this planning effort was similar to designing an elaborate and expensive bridge that reached only halfway across a river. Such a gap is perhaps excusable in planning a war, if fighting offers the only alternative to national extinction. In that case, heroic self-defense not only serves transcendental goals (to go down fighting rather than to surrender) but also provides an opportunity for a saving miracle to intervene. When Finland fought alone against the Soviet Union in 1939, the Finnish military staff could not have offered a plan for terminating the war. Finland's stubborn resistance created the opportunity for the "miracle" to occur in the form of an intervention by other powers. As a result, this militarily hopeless undertaking—based on a plan without an ending—ended as the war that saved Finland's independence.

The other extreme, in contrast to the fight for survival against overwhelming odds, is a conflict where it is clear from the outset that the enemy can be overpowered and will receive no outside support to prevent his defeat. Here military calculations by themselves can, for once, cope with the question of how to end the war. The military design in such a case can aim at the annihilation or capture of all the enemy's forces. Examples of such armed conflicts are the American-Indian wars in the early nineteenth century, the suppression of the Hungarian revolution in 1956 by Soviet forces, and the defeat of the Tibetan independence movement in 1959 by Communist China. No negotiation is needed to bring fighting of this nature to a close. Of course, even with such a capability, a conquering nation may seek to reduce the cost and risks of victory; that is, it may try to induce enemy forces to surrender rather than to fight to the last man, and be careful to avoid provoking the intervention of other powers.

Not only military leaders are sometimes guilty of designing wars as if they had to build a bridge that spans only half a river. Civilian leaders, too, may order the initiation of a military campaign without being troubled by the fact that they have no plan for bringing their war to a close.

In 1956 Prime Minister Anthony Eden convinced himself that Britain's foreign interests would be acutely threatened if Colonel Nasser retained his control of the Egyptian government. Eden saw in Nasser a second Hitler and decided that, unlike in the 1930s, there was to be no appeasement. To remove Nasser, in highly secret sessions with the French Prime Minister he planned an attack on Egypt by British and French forces, coordinated with the Israeli attack against Sinai.

Eden and the French leaders gave a great deal of thought to the initial military operations, considering where the British and

French forces should land and how to time the air strikes, and ordering their military planners to ensure that casualties stayed very low. They also carefully designed various public statements and diplomatic maneuvers in the United Nations, particularly to avoid giving the impression of collusion with Israel. To comply with these political constraints for the initiation of the war, the military planners were ordered to choose Port Said as the landing site, instead of Alexandria which they favored. Alexandria would have offered better access to Cairo, a highly important consideration for ending the war. (The war plan did envisage, as a last step, an attack from Suez westward to Cairo.) [4]

Prime Minister Eden, with all the careful attention he bestowed on reducing the costs and risk of the war's beginning, neglected to plan for an ending that would have accomplished his war aims. How could the landing in the Suez Canal area bring about a situation in Egypt that would have resulted in the overthrow of Nasser? (General Beaufre, who headed the French planning group for the Suez campaign, wrote later: "Even had we taken Ismailiya and Suez, the ultimate result would have been the same—we should have been forced to evacuate the Canal Zone, as we were to evacuate Port Said.") [5] If British and French forces would have had to march on Cairo to depose Nasser, how much time would this have required and what losses and risks would have been incurred? Would the new Egyptian government have been able to stay in power short of a prolonged stationing of British and French troops in Egypt?

In deciding whether or not to initiate hostilities, statesmen may attempt to weigh the risks and costs of avoiding war, on the one hand, against the dangers and possible gains of war, on the other. In evaluating the hazards of peace, statesmen are likely to consider how dangerous it might be militarily to let a potential aggressor become stronger, whether there might be adverse political consequences in the case of political appeasement, and how painful might be the concessions needed to prevent hostilities from breaking out. These costs and risks, of course, can be genuinely high. However, in weighing against them the prospects of war, most statesmen will base their evaluations on the war plans prepared by their military experts and on the estimated losses and chances of success anticipated in these plans. Since war plans tend to cover only the first act, the national leadership, in opting for war, will in fact be choosing a plan without an ending.

AS FIGHTING CONTINUES ITS PURPOSES CHANGE

The outbreak of World War I can largely be attributed to the fact that the major European powers misjudged how their mobilization schemes would interact. To be sure, several of the governments harbored ambitions for territorial aggrandizement, and these, too, played a role in pushing the nations toward war. But in July and August, 1914, the primary motivation for the precipitous decisions to mobilize and to launch attacks was the fear of each power that by waiting it would enable the enemy to strike a decisive blow first. This is the reason why the beginning of World War I provides such somber warning for the present era.

Yet, just as important a warning can be seen in the failure to stop the fighting when it became apparent that the premises on the basis of which the nations had entered hostilities were mistaken. Within the first few months all the major powers could see that their military calculations had been wrong. The reciprocal rush into mobilization and war, supposed to preempt the enemy's preemption, had merely locked the powers into a chain of indecisive battles.

Germany's plans, in particular, went astray. After the battle of the Marne and the failure of the Schlieffen Plan, the German government began to fear a war of exhaustion. The German military chief, General Falkenhayn, concluded in November, 1914, that the war could not be won on two fronts and favored a separate peace with Russia. The German Chancellor, Bethmann-Hollweg, agreed and sent a report to the Foreign Office:

Since the defeat of France during the first war period has failed, and in view of the development of our military operations in the West during the current second period of the war, I, too, must doubt whether a military defeat of our enemies remains possible as long as the Triple Entente holds together.[6]

Yet the German government as a whole rejected this conclusion. The Germans held to their initial war aims, which envisaged territorial gains both east and west, while their enemies just as stubbornly pursued their far-flung objectives. As a result, World War I continued for another four years with disastrous consequences for all the major belligerents of 1914: the Austrian Empire and Tsarist Russia disappeared forever, the German Imperial regime was destroyed and German territory severely reduced, and the French and British Empires—although on the winning side in the end—emerged irreparably weakened. Governments on both sides contributed to their own demise by acting as if, once their nation was engaged in war, there could be no substitute for victory.

Several processes help to explain why governments find it so difficult to back out of a war, even if they discover that in entering it they have made a terrible mistake.

Nations on both sides in a war tend to seek a peace settlement that will bring greater and more lasting security than existed before the fighting broke out. In peacetime, nations manage to live with unresolved conflicts and even tolerate the risk that some of these conflicts might lead to war. But once two countries are at war, this tolerance suddenly vanishes. Hence, governments usually make more stringent demands of a settlement for ending a war than they imposed upon the relationship with the same adversary during the prewar period.

Throughout the 1930s the British government bent every effort to appease Nazi Germany. Yet, as soon as World War II broke out, the British government was determined to fight for the elimination of Hitler's regime. It rejected any thought of a compromise peace, even while Neville Chamberlain—the leading appeaser of the 1930s—was still Prime Minister, and while Hitler, after he had defeated Poland, actually hinted that he wanted a settlement. Of course, after the war had started in 1939, the British government was clearly right, from the point of view of England's long-term national interest, in not reverting to a policy of appeasement, whatever the faults and merits of that policy before the war.

In World War I, by contrast, both sides would have fared better had they been willing to return to the situation that prevailed up to 1914 or even to concede certain gains to the opponent. Instead, they were seized by the notion that the "threat" from the enemy side had to be removed "once and for all" and that they ought to fight on "for a lasting peace." Each side wanted to make territorial gains that would prevent the enemy from fighting another war.[7]

In a prolonged struggle between enemies of roughly equal strength, the aim of preventing future wars sometimes comes to overshadow the initial reasons why the nations chose to fight. Such a preventive prolongation of a war raises several serious questions. In the first place—and this is of capital importance—each side risks losing the ongoing war while trying to improve its position for some possible future conflict. (This happened to Germany in World War I.)

Second, fighting for peace terms that will strengthen the nation's military position at the expense of its enemy is not the only way of reducing the threat of a future conflict. An alternative route, less costly and perhaps more reliable, may be reconciliation; but territorial annexations to improve military security will, of course, work against this solution. Reconciliation was perhaps a more acceptable war aim in the nineteenth century than in the twentieth. Bismarck maintained that the primary task of his government was to reconcile, as much as possible, the nations with which Germany had fought wars. He felt he had succeeded with Austria. But in the peace settlement that ended the Franco-Prussian War (which Germany won), the German military authorities demanded certain annexations of French territory, while Bismarck favored drawing the new boundary along linguistic lines. The incorporation of French-speaking Lorraine, he anticipated, would aggravate the task of creating a lasting peace.[8]

Third, not only can reconciliation be an alternative route toward a more lasting peace, but today's enemy may become a future ally, while today's ally may pose a future threat. Statesmen who become absorbed in the problems of fighting an ongoing war tend to view the present enemy as the permanent enemy. A remarkable exchange between President Wilson and British Prime Minister Balfour illustrates this proclivity. After America's entry into World War I, in a private meeting in Washington, the two men discussed all the major war aims of their nations, now joined in alliance. Balfour objected to Wilson's view that Poland should be big and powerful enough to serve as a buffer state between Germany and Russia. He feared that this would hurt France more than Germany, for it would prevent Russia from coming to France's aid in the event of an attack by Germany. Wilson thought that they had to take into consideration the Russia of fifty years from then rather than the Russia of that day. While they might hope, he argued, that Russia "would continue democratic and cease to be aggressive, yet if the contrary happened, Russia would be the menace to Europe and not Germany."[9] It is ironic that Wilson's foresight was correct, but for too distant a future. Before this reversal in the role of Russia and Germany took place, Germany was again a threat to Britain and France (a weaker Poland, however, might not have increased the chances of a Franco-Russian alliance against Germany in the summer of 1939).

Thus, by prolonging a war to obtain a settlement that seems more secure than the prewar situation, nations can be grievously mistaken for several reasons. They may go down in defeat while fighting for a "lasting peace." They may render more difficult the task of future reconciliation with the enemy. And they may mistakenly assume that the present enemy must be their permanent

enemy. Related to these—frequently mistaken—reasons for prolonging a war is the fact that, after months or years of fighting,
many citizens will come to feel that the outcome of the war must
"justify" past sacrifices. Given this mood, the higher the costs incurred, the more important will it seem that the peace terms be
viewed as achievement of a victory or at least a significant gain.

Prolonged fighting, of course, also affects a nation's war aims
in the opposite direction. As the costs of the war mount, people
become less willing to incur future sacrifices merely to "justify"
past ones. Such an increasing war-weariness descended on the
French nation as it fought the Algerian rebels. Initially, there was
widespread support for keeping Algeria as part of France. Thus in
1954, a few days after fighting had broken out, to a suggestion
that Paris should negotiate, Minister of Interior François Mitter-
and (who was politically rather to the left) could respond that with
regard to the rebellion in "the Algerian departments, the only negotiation is war." Seven and a half years later, de Gaulle granted
the Algerian rebels their every major demand (including some
they had hardly thought of when starting to fight, such as the Sa-
hara), and was supported in this stand by an overwhelming majority of the French population.

Some wars do not have a clear-cut ending. Temporary lulls in
the fighting, or cease-fires, may alternate with high levels of violence. In such a situation governments face less clearly the choice
as to whether to prolong the fighting in the hope of securing a
"lasting peace," or whether to accept a less satisfactory settlement
so as to end the war quickly. The net result of such an ambivalent
situation is hard to predict. On the one hand, a lull in the fighting
without a more definite settlement is likely to keep tensions at a
high pitch, since both sides must remain poised to resume full-
scale warfare at any moment. On the other hand, the lull may last
longer than both sides expected and gradually permit mutual an-
tagonisms to decline.

Changes in the internal structure of governments furnish a

further and particularly important reason why wars are easier to start than to stop. When a nation becomes engaged in a major war, a new set of men and new government agencies often move into the center of power. As "diplomacy breaks down" at the beginning of hostilities, the roles of foreign ministries in dealing with the enemy are much diminished. Officials in the foreign ministries may still write background papers and plan for possible future peace settlements, but the influence that comes with day-to-day decisions is transferred to military staffs. At the very moment that the diplomats are being expelled from the enemy capitals, the military leaders come to command a vastly increased segment of national resources.

This shift in political influence means that the governments on both sides in a war will be concerned primarily with their current military efforts. Once the fighting has started, the military services have a job to do. Their organizational drive and professional dedication may tend to keep the belligerents locked in conflict, particularly since it takes two sides to make peace. If one side, having suffered setbacks, should be eager to return to the conditions that prevailed before the fighting started, the other side is likely to have become relatively stronger—or to think that it has—and hence may elevate its war aims.

It would be an egregious distortion, however, to make the generalization that the military are universally less in favor of peace than diplomats. Indeed, in many a crisis that could have led to war (or did so), military leaders advised greater caution in the councils of their government than did their civilian counterparts. And in several instances, as we shall see, military leaders took the initiative in ending a war, while diplomats and politicians hesitated or even wanted the fighting to go on.

MEANS BECOME ENDS

As the fighting continues, the nations involved in it must reconsider—or consider for the first time—how the war might be ended and how their aims might be accomplished. One should not take for granted, however, that the purposes for which wars are fought can only be realized beyond the fighting. Two views may be held about this.

According to the first view, nations fight wars in pursuit of postwar objectives. That is to say, the government on each side has a plan linking its entire war effort to some well-articulated war aims: mobilization serves the fighting; the fighting serves to subjugate or ward off the enemy; and this in turn serves to bring about the desired peace settlement. War aims will be consciously pursued from beginning to end, even though they might have to be modified because of adversities during the fighting.

According to the second view, however, governments do not decide on the basis of such a grand calculus. Instead, various agencies and individuals in each nation compete in shaping policy, while pursuing their own interests and relying on their divergent estimates of friendly and enemy strengths. Everyone among those involved in guiding the nation through a war may well act rationally in pursuit of his special objectives and in line with his particular image of the world. But the government as a whole cannot be pictured as deciding "rationally." Sometimes this second view stresses a single partisan interest as being the driving force in the decisions on war and peace; sometimes the tableau is richer. The "munitions makers" want to sell their goods, the officer corps wants to enhance its influence and standing, opposition groups want to gain power by exploiting the people's weariness of the war, generals want to vindicate their earlier decisions, and diplomats want to promote themselves as peacemakers.

According to this second view, war serves many purposes, and these purposes are not only accomplished as the fighting ends but are also realized by the war effort itself and the preparations for it. Naturally, those involved will then focus most of their attention on the means, rather than on how the over-all effort will accomplish some ultimate national ends. More absorbing than the final outcome are the perfection of the tools and the mastery of the components and maneuvers that form part of the undertaking. Increasing the performance of a new fighter aircraft or acquiring the latest nuclear engines for a naval carrier is most often the kind of objective about which military services exert their energies in peacetime. The question of how these implements will serve to terminate future wars may be considered only in passing, since the means are desired for themselves. When it comes to actual fighting, the scores that count are, for instance, the number of enemy units destroyed, square miles of territory gained, and other successes or failures in separate battles. Where such an attitude prevails, professional military men would consider it unusual, if not somewhat improper, to ask whether these "mid-game" successes will improve the ending.

According to the first view, on the other hand, governments look at the whole situation in deciding when and how to terminate a war. That is, each warring nation is seen as being guided by a broad analysis consisting of several component estimates: (1) the kind of peace terms that might be attainable without further fighting; (2) how the nation's military situation would be strengthened (or weakened) if fighting continued, and how this would in turn affect the final outcome; and (3) the costs of further fighting and whether these costs would be justified by the prospects of improving the outcome.

Reality seems to be richer than either of these views by itself would suggest. In deciding how to end a war, the top government leaders usually do not altogether lack a broad view; that is, it does make sense—within limitations—to talk of a national decision and of national objectives. Most senior officials are guided by a mixture of considerations: on the one hand, unselfish conceptions of the national interest; on the other—less consciously perhaps— partisan or personal motivations. If one emphasizes only the personal and factional element, or only the unitary view of the nation's interest, one does not do justice to reality.[10]

According to the unitary view of the nation's interest, the question of terminating a war ought to arise as soon as the war has begun, or, indeed, in any advance planning. If nations were such rational, unitary entities, *both* belligerents would have to expect, when the fighting starts, that somehow the terms upon which they will end the war will be better than those upon which they could still avert it. Before the war begins, restraint or appeasement might be rejected in the hope or belief that the enemy is bluffing, or that he will be deterred. But once the fighting has begun, according to this unitary view, it should continue only if, and so long as, both sides expect more from a later ending of the war than from an immediate settlement (be it a more or less limited gain, a stalemate, or a surrender).

In fact, fighting often continues long past the point where a "rational" calculation would indicate that the war should be ended —ended, perhaps, even at the price of major concessions. Government leaders often fail to explore alternatives to the policies to which they became committed, and they may even unconsciously distort what they know so as to leave their past predictions undisturbed. For instance, a government may continue fighting a war in order to move toward a hazily perceived "compromise," while failing to take into account the intentions and capabilities of the enemy with whom the compromise would have to be reached.

After the initial battles, a great deal more is known about the enemy's relative strength than before the war. Hence, an approach that rationally considered only the interests of the nation as a whole would call for a reevaluation of the decision to fight, since —in rational decision-making—new information leads to new (or reaffirmed) choices. Yet a government rarely reverses itself after the first campaigns of a war, treating the decision to fight as a mistaken choice that ought to be rectified.

If the decision to end a war were simply to spring from a rational calculation about gains and losses for the nation as a whole, it should be no harder to get out of a war than to get into one.

TWO

THE FOG OF MILITARY
ESTIMATES

A great part of the information obtained in war is contradictory, a still greater part is false, and by far the greatest part is subject to considerable uncertainty.
—KARL VON CLAUSEWITZ

MODERN GOVERNMENTS, in guiding their country through a war, can draw on vast masses of data and avail themselves of day-to-day, or even hour-by-hour, reports from their far-flung intelligence systems. Nonetheless, some of their key decisions have to be taken in the face of great uncertainties. But government leaders frequently fail to acknowledge these uncertainties or to take them into account in their decisions. Instead, they often implicitly assume answers to questions that they have never examined.

For any war effort—offensive or defensive—that is supposed to serve long-term national objectives, the most essential question is how the enemy might be forced to surrender, or failing that, what sort of bargain might be struck with him to terminate the war. This question combines both the military and the political realms, in the sense that not only the military contest but also do-

mestic and foreign policy developments will determine the outcome. As we have seen, this aspect tends to receive short shrift in
war plans. Sometimes it is ignored entirely.

Next in importance is the question of how one's over-all military strength will compare with that of the enemy, taking into account the resources that might be mobilized on both sides and the
various potential limitations, or escalations, that might be imposed
on the conflict. Answers to this question must be based on military estimates. The data that military intelligence ordinarily produces, however, provide only the component elements, such as estimates of the enemy's order of battle, territory gained, munitions
expended, future logistic requirements, and losses in men and
matériel among friend and foe.

More aggregated estimates are needed to make forecasts
about the course of the war. And being more aggregated, these estimates are also more abstract, and hence harder to construct and
harder to agree on. Government bureaucracies are ill equipped to
develop such evaluations.

Military staffs devote most of their work to details of battles
and campaigns and to daily operational activities. The amount of
time left to think about and plan the war as a whole is minute in
comparison. Thousands of man-years are absorbed by calculations
on specific military operations or pieces of equipment, and by intelligence estimates of enemy strength in this or that local area.
Usually very few military officers or civilian analysts are given the
time and opportunity to put all these pieces together and to prepare estimates that bear directly on the over-all strategy and that
will help to show how the entire undertaking might be brought to
a satisfactory end.

On the other hand, top government leaders—who must ar-
rive at the broader, "political" judgments—often are insufficiently
conversant with the hard facts of the military domain. This obliv-
iousness to military reality is starkly illustrated by the decisions of
the French government in 1939 and 1940. Unable to appreciate

the military balance of forces, French prime ministers at that time sponsored the strangest schemes for fighting Germany. Likewise, Hitler declared war against the United States without having looked at, or even requested, an evaluation of how German forces could fight America. The decisions by Nasser and King Hussein that brought on the Six-Day War in 1967 were also apparently taken without the benefit of military estimates of how the war might end.

Government leaders have sometimes tried to probe more deeply into the detailed estimates of their military staffs and to feed them the kinds of questions that are relevant for the top decision-maker. The trenchant memos that Winston Churchill sent to many parts of his intelligence services during World War II come to mind. But if this probing into detailed estimates is done at the expense of broader evaluations, the top leader will lose sight of the forest in his scrutiny of the trees. Who is left to guide the nation through the war and to keep the costs of fighting in line with the long-term benefits, if the top leader uses his precious time to pore over maps of field commanders or to decide whether or not to bomb that power plant?

Even if military staffs and the political leadership were able ideally to divide their responsibilities, each focusing on the right level of detail, major intellectual difficulties would still remain in generating the estimates for deciding how to end a war. The more essential it is to find "a substitute for victory," the greater these difficulties. A battle won should count on the plus side only if it fits into a larger design for ending the war on favorable terms; otherwise it might have as disastrous consequences for the winner as did the battle the Japanese won at Pearl Harbor. Some common ways of judging military success—such as territory gained or improvements in current force ratios—are insufficient or even misleading indicators for guiding the conduct of war.

Traditional military estimates, like those reflected in a commander's war map with its pins and blue and red lines, are uncertain enough. The fog of uncertainty thickens indeed if the estimates are broadened, as they must be to plan how to terminate a war in a desirable fashion—whether a war in being or one that might start in the future.

While governments thus find it extraordinarily difficult to calculate beforehand how a war might end, they can at least consider the key uncertainties, so that they may weigh the risks of initiating (or prolonging) a war against the risks of settling with the enemy. If a statesman decides to go to war, or to reject opportunities for ending an ongoing war, he must somehow assume that fighting— or fighting on—will improve the outcome. Far too often, this assumption receives no analysis. Whether or not it is warranted depends on several broader considerations:

the potential on both sides for mobilizing men and resources;

the possibilities of outside help, to the enemy and to one's own side; and

the pressures of the war's violence on public morale and on government views of the national welfare; in other words, the impact of the costs of war—casualties, human suffering, and economic losses, both past and expected—on popular support for the war effort and on possible government decisions about ending it.

Without these additional considerations, strategy remains divorced from its objective, and all the battles and campaigns are a succession of means unrelated to their end.

WHAT RESOURCES WILL BE USED?

The potential for mobilizing men and matériel is determined on each side by the amount of resources that can be made useful for war and by the extent to which the government and people are willing to allocate their assets to war. In both world wars, Germany underrated not only the American capacity for industrial production but also the willingness of the American people to devote resources to fighting the war. The extent to which resources will actually be mobilized is of course far more difficult to estimate accurately than a country's capacity for production. But productive capacity, too, may be grossly miscalculated. In 1944-45, the majority view in the American intelligence community overrated Japan's productive capacity. Had Japan's surrender not intervened, this exaggeration could have led to costly mistakes in the American plans for ending the war.[1]

The extent to which economic resources will be made available for war is so difficult to predict because it depends on states of mind in the top leadership and in the population at large. Dictatorships may be as reluctant as democratic governments to press for full mobilization. Adolf Hitler, for example, consciously limited his war effort until the last year of World War II. As late as 1943 he still opposed the conscription of German women: "The sacrifice of our most cherished ideals is too great a price." [2] His squeamishness disappeared only when Allied forces began to converge on Germany.

On the other hand, in World War I, the parliamentary governments of England and France were able and willing ruthlessly to demand the last sacrifice from their people. These sacrifices served war aims that went far beyond the enemy's expulsion from the territories he had invaded. The British government kept sending new troops into the murderous battles in Flanders without relinquishing any of its far-flung imperial war aims that had so little to do with the freedom and security of England: Constantinople and Courland had to be handed to the Tsar, Mesopotamia had to come under British control, German-speaking provinces of Austria had to go to Italy. Similarly, the French government presided over the decimation of France's manhood and the depletion of her wealth without ever considering peace at the price, say, of letting Germany keep German-speaking Alsace, while recovering only Lorraine. (Alsace, after all, had been German territory until France acquired it in 1648 after the Thirty Years' War, and it had again been under German rule for forty-four years prior to World War I.)

A great contrast appears some forty years later in the willingness—and ability—of the French government to demand sacrifices for the war in Algeria. Only a small fraction of the mobilizable resources were ever devoted to the goal of keeping Algeria French. And yet, when the rebellion broke out in the fall of 1954, most Frenchmen still considered Algeria just as French as, say, Corsica, or as Alsace during World War I.

In a war where both sides have mobilized only a fraction of their available resources, the role of the remaining potential is particularly hard to assess. Yet, in terminating such a "limited war," the confrontation of latent strength might be as important as the confrontation of actual strength on the battle front. If our enemy refuses to settle on terms acceptable to us, we can mobilize additional forces to overcome his *current* strength. But how should we estimate his *future* strength? Will he be unwilling to mobilize further, or will he match our increase and thus restore the stalemate? We can collect statistics regarding the number of men, airfields, tanks, ships, and so forth, to indicate current force ratios; and we can estimate the enemy's maximum potential for war production and military manpower. But neither set of figures is likely to tell us how far, at the war's end, the enemy will have gone in transforming his potential into actual strength.

In the spring and summer of 1952, after the Korean armistice negotiations had dragged on for a whole year, the American military planners wondered how the war could be terminated. A memorandum prepared by the planning staff of the Joint Chiefs in May, 1952, stated: "At the present time we are faced with a set of conditions in Korea which preclude, from the military point of view, a conclusion which can be termed satisfactory." This memorandum estimated that it would take at least a year to increase U.S. military strength so as to ensure Communist acceptance of an armistice on United Nations terms. But the question whether such an increase would not be canceled by an increase on the Communist side was apparently not examined.

Another American intelligence estimate for the Korean war, in the summer of 1952, concluded that the Communist side was content to rest on its increased defensive strength, confident of its ability to wait out the United Nations Command. Unless the United Nations Command mounted a major offensive and "broadened" the geographical limits of the war, the authors of this estimate believed, sufficient military pressure could not be applied against the enemy to bring about a swift conclusion of the war.[3] Again, this estimate remained silent about the other side of the coin: it did not examine to what extent the enemy might also mobilize more forces and be able to respond within the "broadened" geographical limits as well.

HOW MUCH OUTSIDE HELP?

The difficulty of estimating the extent to which the enemy will mobilize more of *his* resources is compounded by the uncertainty regarding outside help (or hindrance). One's own mobilization potential and that of the enemy are not isolated from the rest of the world. On the one hand, they may be augmented by friendly powers and future allies; on the other hand, some military forces may have to be reserved for further contingencies. External aid might enlarge the resources available for the war effort, real or imagined threats from other powers might draw forces to another front, and new conflicts between third parties in other theaters might radically alter the ambitions and restraints of the original belligerents. Only for an all-out conflict between superpowers can the military potential of the outside world be ignored.

In June, 1940, when the Pétain government was about to conclude an armistice with Hitler, the French military had to evaluate the prospects for offering further resistance from the French possessions overseas. One of the key questions was whether or not the Germans could overwhelm the French forces in North Africa.

In light of the strength of the powers at war at that moment, the prospects for a continued French effort based on North Africa seemed bleak. On one side stood the victorious German army and air force supported by the Italians; on the other side were the French forces that happened to be located overseas or the remnants that could still be moved there, plus the French and British navies, the small British air force and army stationed in England, and the modest support that members of the British Commonwealth could have offered. At that time, it was reasonable to expect that Franco's Spain would have made common cause with Germany in an attack on Gibraltar. From Gibraltar, the German forces could have moved into Spanish North Africa and attacked the French forces in Algeria, perhaps supported by an Italian attack from Libya against Tunisia.

But the existing forces and the mobilization potential of the two belligerent sides were not all that mattered. From hindsight it is now easy to see that the chances for a continued French war effort in North Africa would have been helped by Hitler's deep distrust and hatred of the Soviet Union, which led to his decision a few months later to plan his attack in the East. During the first half of 1940, however, the French leaders did not see a potential ally in the Soviet forces. On the contrary, they not only underestimated Soviet military strength grossly but misread the future prospects of the war to such an extent that, up to the German offensive in the West. they kept a plan alive for a gratuitous Anglo-French attack on the Soviet Union with the ostensible purpose of ending Russian oil deliveries to Germany.

The weaker the country, the more ought its military planning to take account of the possibilities for outside help, both for itself and for its enemy. But government leaders are easily absorbed by what they see just across their border; and if their ambitions are focused on the province next door, their military vista may fail to extend much farther.

In 1950 the North Korean leaders probably focused almost exclusively on the strength of the South Korean forces relative to their own, and calculated—quite correctly—that they could easily defeat the South Koreans. If the Arab-Israeli confrontation continues through the 1970s, one can imagine that either side may someday become trapped in a similar overemphasis on the local balance of power. A particularly dramatic example of this myopia is offered by Finland's decision in 1941 to join in Hitler's attack on Russia.

In 1939–40 the Finns had barely saved their national existence by managing—at the price of painful territorial concessions —to end their brave and lonely fight against the massive Soviet attack. But fifteen months later, when Hitler's armies marched into the Soviet Union, the Finnish government decided the time had come to break its peace treaty with Moscow and to recapture lost territory by joining in Hitler's attack. Today one can only wonder why the Finns tied their country's fate to Hitler's conduct of the war before more of the military uncertainties could be resolved. They could see, after all, that Hitler had failed so far to defeat England or to cow her into a new appeasement, that America offered substantial support to England and might yet come in on her side, and that Mussolini's armies had already been beaten on several fronts.

To be sure, strong forces pushed Finland to break her peace treaty with the Soviet Union. During 1940, Finland still felt gravely threatened by the Russians (for good reason: but for Hitler's veto, the Russians would probably have attacked Finland again late in 1940 and defeated and occupied her completely). As a result of this threat, the Finns sought support where they could find it and thus gradually became entangled with the Germans. Early in 1941, Finnish military leaders were tempted by the opportunities that would be opened by a German attack on Russia; but they remained worried about the possibility of another Nazi-

Soviet deal. "There was just one thing they never seemed to have considered at all. That was the possibility that if the war did come, the USSR might win." [4]

Once the Finns had entered World War II on the German side, the success of their own offensive (by which they more than recaptured the territory Stalin had taken from them) became secondary. What mattered most for Finland was the fate of Hitler's offensive in Russia and Hitler's war against Britain and the United States. On February 3, 1943, the day after the German forces had surrendered at Stalingrad, the President of Finland came to see Finland's military leader, Marshal Mannerheim. The two agreed, after discussing the military outlook, that Finland had to make peace with the Soviet Union. On February 9 the head of Finnish military intelligence gave the parliament a report of the unfavorable military prospects and told its members that they had better face the need for making peace with Russia.[5] From then on, the military estimates could only reinforce this conclusion. Even though the Finnish army had not suffered any defeats, the outcome of the final round could no longer be in doubt. There was practically no chance for a compromise peace between Germany and the Allies. By prolonging the war, Finland was traveling down a road that could only lead to total and final disaster. Yet it still took the Finnish government a year and a half before it found itself able to accept the Russian peace terms.

In Finland's war with Russia on the side of Nazi Germany, the possibilities for outside help on both sides were clear. Nothing remained in abeyance that might shift the balance, for after 1941 all the major powers were aligned on one side or the other. And all of the powers involved committed most of their military and industrial strength to the war. Even though the civilian economies were mobilized with varying intensities, none of the powers conducted the fighting as a limited conflict where the main forces had to be saved for another contingency.

However, in a war where major powers participate with only a fraction of their mobilizable resources, the relevance of estimates regarding the global balance of forces is much less clear. To decide how and when a war should be ended one must try to estimate the military situation up to the final round of fighting. But how can one tell at the beginning of a limited war what the final round will be? How wide should the circle be drawn to estimate the line-up of forces? How many steps in possible escalations and expansions should one calculate to determine one's strategy for fighting and to guide one's bargaining to end it?

In the Korean war, for example, the deadlocked truce negotiations must have led both sides to look beyond the local balance of forces. Vast resources could still have been brought to bear on the fighting by allies on either side. While possibilities for such escalations were easy to see, their bearing on the truce negotiations, nonetheless, could not be calculated with any precision. That China could have thrown in more forces was obvious, just as it was evident that the United States had much more military power to commit. Was the outcome of a conventional war between the United States and China the important calculation, or the outcome of a nuclear attack on China, or the outcome of such an attack on China and the Soviet Union together? If the military calculations were to be extended that far, perhaps they should have been stretched even further to assess another round of fighting that might have followed such a nuclear exchange in the early 1950s, such as some "broken-back" warfare by the Red Army against Western Europe.

The prospects for increased outside involvement in a war can be judged in many ways. How likely is each larger round and how might it end? Is the enemy emboldened by the expectation that his allies will always bail him out, or is he afraid to lose their support and hence inclined to make peace? It is easy to see that senior government officials can disagree on these questions.

WILL THE COST COERCE THE ENEMY?

Further difficulties arise in estimating how military operations will contribute to ending a war. The violence of war has a dual impact. It not only serves to overwhelm the enemy's military forces, but it can also make him eager to settle by hurting his population and economy. This second effect has to be figured into military evaluations as much as the first.

Thus, bombing of cities might not only reduce the capacity to produce weapons but also undermine civilian morale and hence lower the war effort indirectly or increase homefront pressure on the government to sue for peace. (The bombing campaigns in World War II were aimed at both of these effects.) A blockade might not only interdict military supplies but also cause food shortages leading to popular unrest. (This happened in World War I.) And continuing high casualties at the front will not only decimate the troops but might also compel the government to seek a faster end to the fighting—for humanitarian reasons, for considerations of national welfare, or because of public pressures. While these gross secondary effects can be easily expected, it is extraordinarily difficult to predict their strength and net impact.

Each side in a war has to decide whether it should merely seek to fight the enemy's military forces, or whether it also wants to wage campaigns designed primarily to hurt the enemy's population or economy. In addition, some resources might have to be allocated to keep the enemy from inflicting too much pain to one's own side. (In nuclear war, these distinctions are posed strictly: nuclear deterrence is almost exclusively a threat to hurt the enemy's homeland, not to ward off or overwhelm his military forces.)

Some major campaigns in past wars served solely the purpose of coercing the enemy by hurting his homeland. For instance, in World War I, the German government decided—after lengthy internal debate—to start unlimited submarine warfare in order to cripple the English economy and thus compel England to sue for peace. As it turned out, the German government made a cata-strophic miscalculation.

Mistakes in estimating the coercive effects of warfare are common, though not always so costly for the initiator as was this German error about submarine warfare. An example from World War II is the British decision to bomb the residential and central business areas of German cities. As it turned out, these "area raids" of the British Bomber Command fell far short of the results that had been expected by the British in 1941. They failed in their main purpose of seriously damaging the morale of the German labor force (although in some untraceable way they might have helped to speed up the surrender of German forces in 1945, and they did tie down German resources in air defense). This use of their resources was not the only course open to the British. Even if the British Bomber Command could not have acquired the type of aircraft to attack military targets (such as aircraft factories and oil refineries) in time, the British did have the option of investing less in their bomber force, and using resources elsewhere—for in-stance, on antisubmarine defense.

Prior to the decision to go ahead with the bombing cam-paign, British defense analysts did in fact debate whether the pro-posed attacks against German cities would be effective. This debate has now acquired some fame because of C. P. Snow's ver-sion of the controversy between Professor Frederick Lindemann and Sir Henry Tizard, both of whom were scientific advisers to the British government. What is of interest here is that everything centered on the calculations of the number of planes that could be put to use, the number that could reach the target, and the physi-cal effects of the bombs dropped. Those criticizing the estimates never challenged the one crucial sentence in the memorandum by Lindemann (who was close to Churchill) recommending the area attacks. This sentence was the only link between the estimated physical effects and the desired results of this coercive campaign. "Investigation *seems* to show," the memorandum argued, "that having one's house demolished is most damaging to morale. People *seem* to mind it more than having their friends or even relatives killed. . . . There *seems* little doubt that this would break the spirit of the people." (Italics, needless to say, added.) As it turned out, it was this judgment particularly that was mistaken.[6]

On the basis of the German wartime statistics that have since become available, it is easy to show what went wrong. The Lindemann judgment implied that the number of people made homeless was an important criterion for the effectiveness of the bombing strategy, forgetting that people can double up without giving up their jobs as long as a large percentage of a city's houses remain intact. In 1943–44, Berlin lost 40 percent of its dwellings, but the number of workers in essential industries did not decline at all.[7]

A parallel misjudgment was made by the British government in 1938, in studies of the possible effects of German air raids on London and other British cities. These studies estimated that the German air force could inflict high casualties and destruction, and predicted that an attack of such dimensions would constitute a "knockout blow" to England. Quite apart from the question whether the figures were right (the British military analysts, it turns out, made many serious errors in their calculations),[8] nobody paused to examine the judgment that a certain amount of casualties and destruction meant a "knockout blow." This metaphorical expression merely covered over what should have been viewed as a major uncertainty. Metaphors always pose dangerous traps for the policy analyst.

Since it is so difficult to estimate how the deprivations of war will affect the enemy, government leaders simply have to guess. Yet these guesses are critical for decisions on how to terminate a war. And because they must be guesses, based on soft intelligence, they can provide easy opportunities for self-deception. When the news from the battle front is discouraging, a tempting consolation is to assume that the enemy is suffering so badly from the war that he is about to sue for peace. Judgments about the enemy's suffering—unless they are made with mental discipline—can serve to avoid unpleasant but necessary decisions. They can be the excuse for not grappling with the prospect of an indefinite stalemate or even defeat.

In the last year of World War I, Germany's chief strategist, General Ludendorff, set his hopes on a massive spring offensive against the Allied forces in the West. As he explained later on, this offensive "was to make the enemy ready to accept peace." On August 8, 1918, however, when several German divisions were defeated and the Allies began to advance, his hopes were dashed.

Six days later, the Kaiser attended a war council at General Headquarters in Spa. The single most critical forecast at this top-level conference was that the enemy would be coerced by the costs and pain that the war inflicted on his homeland. Precisely because they were able to delude themselves as to these coercive effects did Germany's top leaders fail to seek a compromise peace in the summer, before their setback turned into a definitive defeat in the fall.

During the first part of this war council, the German Foreign Minister started the deliberations on the right track. He began by stressing that the enemy was "more confident of victory and more willing to fight than ever." The chief cause, the minister explained, was the conviction of the Allied Powers that "their comparatively inexhaustible reserves of men, raw materials and industrial production must crush the Central Powers, with the help of time alone."

As the discussion continued, however, the war council seemed to forget the warning of the Foreign Minister. The German leaders saw no need for haste in making peace, for they estimated that their "stalwart" nation had an edge in willingness to bear the costs and pain of further warfare. The Kaiser, while "agreeing" with this evaluation of the enemy's strength and self-confidence, nevertheless pointed out that the enemy was suffering, too; many of his men were being killed, his industry was beginning to lie idle for lack of raw materials, and food was becoming scarce: "This year's harvest in England is poor; her [shipping] tonnage is diminishing daily," the Kaiser elaborated; "perhaps, as a result of this shortage, England will gradually turn towards peace." [9]

"Perhaps!" Meanwhile, Germany's own allies not only "gradually turned towards peace" but actually made peace (Bulgaria) or simply collapsed (Austria), Germany's front lines dissolved, her sailors mutinied, the Kaiser had to flee, and the Allies could dictate peace terms.

WHAT IS THE ENEMY UP TO?

Of the many pieces of information that would vitally affect judgments about whether, when, and how to seek an end to a war, the enemy's own plans and expectations are of course among the most decisive. Yet, even if a government obtains such knowledge, the top leaders often fail to take advantage of it or distrust it because of apparently contradictory reports.

In the Korean war, when the United Nations forces continued to advance northward, Washington received various warnings that China would intervene should the advance continue toward the Yalu. From hindsight it seems clear that the United States would have been better off had it heeded these warnings and sought to terminate the war before China came in. It could, for instance, have proposed an armistice when the UN forces had reached the narrow neck of the peninsula north of the thirty-eighth parallel. [10]

Information about the plans of other countries often derives from cryptographic intelligence and other forms of eavesdropping. In almost every war in this century, some of the most secret codes were broken. Sometimes each side was able to decipher the other's messages while remaining blissfully confident that its own communications remained secret. These incidents show the remarkable power of modern intelligence services in ferreting out crucial information. And the fact that governments in possession of this information so often failed to exploit it shows the remarkable difficulty of using this kind of knowledge.

One of the most spectacular failures to make use of a broken code occurred prior to the attack on Pearl Harbor, when the American government did not take advantage of its ability to decipher the Japanese code. A successful exploitation of a broken code occurred in World War I, when the British government pushed President Wilson closer to the decision to declare war on Germany by showing him a decoded German telegram which urged Mexico to join with Japan in war against the United States, promising American territory as a reward. At first, however, there was considerable resistance within the British government to the idea of exploiting this telegram, for fear of revealing Britain's knowledge of the German code. Incidentally, in World War I the Germans, too, managed to break British codes and derived some tactical advantage from this information, as did the French by breaking German codes used for military communications.[11]

In World War II, the Germans were able to listen in for a period of time on the telephone conversations between Churchill and Roosevelt by breaking the scrambler code—also without major gains. For instance, in July, 1943, they intercepted discussions regarding the Italian surrender negotiations. And there were several other instances in World War II where the German intelligence service discovered important plans of the Allies while the leadership failed to exploit this knowledge. After Pearl Harbor, the U.S. government continued to decipher Japanese communications and took advantage of the information gained thereby in tactical operations—among them, the battle of Midway.[12]

Then, during July, 1945, the United States intercepted Japanese requests to Moscow for Soviet mediation to end the war. The Soviets stalled in acting on or responding to these requests, but at the Potsdam Conference Stalin at least informed Truman of the gist of the Japanese telegrams. The fact that a few top-level people in the U.S. government had advance and detailed knowledge of this Japanese peace effort permitted the President and his closest advisers to appreciate better the concern within the Japanese government about the Allies' stated demand for "unconditional surrender" and whether this demand required Japan to relinquish the Emperor. For example, Secretary of War Henry Stimson urged the President in a memorandum at the beginning of the Potsdam Conference, by referring specifically to these Japanese peace feelers, to deliver the planned warning to Japan (the Potsdam call for surrender) more promptly. As it was, the American government might have acted more decisively on the basis of the information in these deciphered telegrams. Had the Japanese been told earlier that they could keep the Emperor, the war could have been shortened by a critical week or two. Since only a few top officials in the American government were privy to these messages, they could not be analyzed with the care they deserved.[13]

WHICH ESTIMATES TO BELIEVE?

It often happens in wars that the weaker party makes no attempt to seek peace while its military strength can still influence the enemy, but fights until it has lost all its power to bargain. Various reasons explain this self-destructive perseverance. The loser may feel that any attainable peace—short of "victory"—is about equally bad; or the struggle among factions at home may keep him fixed on his road to disaster. In such situations, government leaders find all kinds of excuses for disregarding military estimates, no matter how compellingly they point toward eventual defeat.

In World War II, for example, after the German armies had suffered their reverses in Russia and the Allies had landed in North Africa, Mussolini at first chose to ignore the military implications for Italy. In December, 1942, Mussolini's chief of military intelligence learned from Admiral Canaris (his German counterpart) about the unfavorable prospects for the German armies in Russia. But when he reported this information to Mussolini, the latter showed no interest in it. In fact, Mussolini asserted his confidence in quick victory.[14]

By the following spring, some of Mussolini's officials forced him to pay more attention to the military prospects. In a meeting on March 10, 1943, his Minister of Transport—at whose urging the meeting had been called—presented grim statistics about the Italian merchant fleet. If present trends continued, the minister showed, only 450,000 tons would be left by June. Mussolini commented that he was "gratified for the clear and explicit exposition which only today, March 10, 1943, is shown in its full reality in contrast to the vague figures which have been hitherto given." After some further discussion of Italy's disastrous situation in transportation, the Minister of Transport audaciously ventured that "it would be interesting to hear, after these comments, who is responsible for the conduct of the war."

Mussolini now had to face the facts or muddy them up. He chose the latter course:

At the worst today we dispose of 600,000 tons of shipping, and with a loss of 100,000 a month in six months we shall be left with only fishing boats. The essential is to protect our patrimony now while preparing to safeguard what we shall have later. Anyway no one can state that our losses will be at the rate of 100,000 a month.[15]

It is easy to see why leading officials within a government can differ so widely about how their country should end a war. Commitments to personal values and domestic political constraints provide the motives; the complexity and uncertainty of military estimates furnish the opportunity. Even if the leaders start out with identical data, they must use intuitive judgments to render these data relevant for the termination of the war. They must decide which data to ignore as trivial and which to interpret as important signals. They must reconcile conflicting evidence. They must amalgamate into a single answer the most diverse indicators: reports from the battlefield, statistics on potential military resources, and impressionistic predictions of how friend and foe will bear the costs and suffering of further fighting.

That is to say, the uncertainties and inaccuracies of military evaluations of ways to end a war are only partly to be blamed on missing data. In large part, they stem from the scope left to intuitive judgment—or gross bias—in putting the data together. In order to assess the fighting strength of nations in the midst of a war one must aggregate far more diverse facts than to evaluate the worth of a large corporation or to calculate a country's balance of payments. Yet, even in these cases, where everything can be measured in dollars and cents, the uncertainties are far greater than laymen realize.[16]

Because of this latitude in fitting the pieces together, political as well as military leaders can suddenly reverse their interpretation of the war. Initially, they may brush aside unfavorable data, and in interpreting uncertainties always stress the optimistic side of the range. Then, after some dramatic setback, they may suddenly see their world in a new and somber light and stress all the pessimistic possibilities. It is as though a promising but somehow mystifying acquaintance on whom one has set great hopes is suddenly discovered to have committed some evil. All the potentially disturbing earlier signs—that one has chosen to explain away or disregard—now seem to fall into place. Indeed, additional evils may now be attributed to the person quite groundlessly. The government leader who realizes at last that his "road to victory" is in fact a road to disaster may underestimate his remaining opportunities.

On August 8, 1918, Ludendorff suddenly abandoned his firm confidence in a German victory after several of his divisions had been defeated, and precipitously demanded that his government ask for an armistice. In World War II, Mussolini realized, more or less consciously, that Italy was defeated when the Allied forces landed in Sicily. (And this change in his outlook may account for the resignation with which he accepted his ouster.) In the Korean war, the initial successes of the Chinese intervention caused the sudden abandonment of the United Nations plan for reunification of Korea and stimulated serious U.S.-British consideration of total withdrawal from the peninsula.

The cliché that it is the last battle that counts begs the question, for only *after* the war can we know which battle was the last one. Besides, part of the time this cliché is wrong. The last battle may not count at all, in that it may leave unaffected the decisions of both sides to end the war as well as the conditions for ending it. For instance, in the Korean war the last battle resulted from a Communist offensive in June, 1953, which had to be fought off with heavy UN casualties. Yet, prior to this battle, all the details of the armistice had been settled and nothing changed in consequence of this final fighting. It was only Korean President Syngman Rhee who still had to be won over to the armistice terms, but American pressure would have probably accomplished this without the last offensive by the Communist forces.

The outcome of a single battle—whether or not it is the last—can bring about the termination of a war in either of two ways. On the one hand, if the leadership of the losing side persists in fighting as long as physically possible and can impose its will on its own troops, the final battle will end the war by overwhelming the loser's last forces. This happened when the battle of Berlin, in April–May, 1945, put an end to Hitler's reign.

On the other hand, if the leadership on the losing side does not want to make its forces fight on till they are all destroyed, or cannot, a single battle may be decisive in bringing an entirely new perspective to the war. In this case, a dramatic setback or some particularly painful loss in men or territory provides the trigger for a complete reevaluation of the military prospects. The temporary success of the Chinese intervention in Korea and the setback suffered by the German armies on August 8, 1918, provide examples. At the same time, the impact of the new appreciation of how the war is going and what it costs may strike the spark for a change in leadership. What counts is the psychological shock from such a battle, rather than the actual military effects on front lines and force ratios.

THREE

PEACE THROUGH ESCALATION?

> *Senator Lyndon B. Johnson: Assume we embrace your*
> *program and suppose that the Chinese were chased back*
> *across the Yalu River, and suppose they refuse to . . .*
> *enter into an agreement . . . what course would you*
> *recommend at that stage?*
>
> *General Douglas MacArthur: I can't quite see the*
> *possibility of the enemy being driven back across the*
> *Yalu and still being in a posture of offensive action. . . .*
>
> *Senator Lyndon B. Johnson: Now suppose we inaugurate*
> *this program. . . . They go back across there, they still*
> *retain large mass formations there, what course are we*
> *going to have to take?*—MACARTHUR HEARINGS, *1951*

IN MANY A WAR, the national leadership sooner or
later must face the fact that as the fighting drags on far beyond
what had been envisaged in the initial war plans, new plans are
required to guide the military effort. The more the leadership fears
the costs and risks of continued fighting, the more urgently will
it seek a design for the war that promises to end the fighting
quickly.

But just as the costs and risks of continuing the war are subject to vast uncertainties, so the terms on which it might be possible to settle with the enemy are frequently obscure. If total subjugation of the enemy is judged to be impossible (or not worth the cost), it will be hard to estimate to what extent further fighting will ensure improved terms for peace. Even if explicit peace terms have been proposed by one or the other side, they often mean little, since further fighting is likely to change them.

In most wars since the Middle Ages, one or both of the belligerents could have committed substantial additional resources or in other ways increased the scale of fighting, but for a variety of reasons chose not to do so. Unless all mobilizable military and economic resources have been engaged, the level of violence can be raised by introducing new weapons, by attacking new targets, or simply by deploying more men and matériel. And larger powers often have the additional option of fighting the enemy on new fronts or extending the war to his allies. Such geographic expansion is a particularly significant possibility for opponents with global capabilities and interests: one side can extend the conflict to a distant quarter of the globe.

Given that the terms of settlement offered to the adversary are usually variable or vague or both, and given that the belligerents can usually increase their military effort, the choices for ending a war are doubly indeterminate. Each side has at its disposal both more carrot and more stick.

Today, it has become fashionable to refer to all the ways in which fighting can be expanded as "escalation." This word—in its present meaning scarcely fifteen years old—has the disadvantage of making us think that expansions and contractions in military effort occur along only one dimension—like rungs on a ladder. Instead of a single escalation sequence, however, there usually are competing alternatives; for instance, troop levels might be increased but air or naval operations reduced, or a new theater of war might be opened but another one abandoned.[1]

To clarify the question of escalation we must recall why governments tend to limit their war effort. Five motives—singly or in combination—offer most of the explanation:

First, governments tend to refrain from escalating a war if they expect that the military gains of increased violence would be canceled by the enemy's counter-escalation or by the intervention of other powers on the side of the enemy.

Second, war efforts tend to be limited by fear that an increase in violence might—through various mechanisms—bring about a further eruptive expansion of the fighting to intolerably costly and destructive levels. This argument against escalation has become widely publicized. If nuclear powers fought each other directly but using only conventional weapons, the dominant motive for limiting the conflict would probably be fear that expansion could lead to a gradual or sudden descent into the abyss of thermonuclear war. Perhaps the only historical experience comparable to what is feared here is furnished by the explosive escalation of the war between Austria and Serbia, which remained a "limited war" only from July 28 to August 31, 1914.

Third, governments may fear that an escalation of fighting would inflict destruction and death on their own territory.

Fourth, governments wish to avoid internal dissension and keep down the economic cost and social dislocation needed to sustain the war effort. The social and economic costs, though, inhibit only certain types of escalation. A new campaign based on large numbers of troops forces the government to draft soldiers, whereas an expansion of the geographic limits of fighting or the introduction of new types of weaponry need not involve the citizenry in mobilization. But in either case, adverse political repercussions at home might overshadow these social and economic costs.

Fifth, military capacity is sometimes left unused in a local war as a reserve for coping with other potentially menacing powers.

Given these motives, governments ought to be—and usually

are—as concerned about prolonging a war as they are about "escalating" its level of violence or its geographic spread. A defect of the "escalation" metaphor is that it leaves out this time dimension. As long as a country still has military strength, simply to prolong the fighting at the existing level is one principal way in which it can put further effort into the war. Alternatively, it can fight at a higher level in the expectation that this will help end the war at an earlier date. That is, prolongation of a war may be an alternative to other "escalations."

Despite such a trade-off, however, there are important limitations to the possibilities for interchanging prolongation of a war for escalations in level of violence or in the geographic spread of fighting. Escalation can occur either gradually or by marked and perhaps large steps, whereas prolongation shows no such steps. The decision to prolong fighting might be taken almost by default, while the decision to introduce some new type of weapon or to attack at a new front must be more deliberate.

Obviously, a country cannot prolong a war without maintaining an effort sufficiently high to prevent total defeat, and it cannot escalate without continuing the fighting long enough to bring its increased effort to bear on the enemy. In World War II, for instance, the French Premier was disappointed that the Finnish Winter War did not last long enough for France to escalate the effort against Nazi Germany by sending troops to Norway. (As it turned out, it was fortunate that this expedition in 1940 did not come off, for it could have led to fighting between the Soviet Union and the Western Powers.)

Whether or not a nation can shorten a war by escalation depends on many factors. If a nation can overwhelm all of the enemy's forces by escalating a war, the fighting will of course be brought to an end—provided that no outside power intervenes to help the defeated country. Short of inflicting such total defeat, successful escalation would have to induce the enemy government to accept the proffered peace terms. The trouble is, the greater the enemy's effort and costs in fighting a war, the more will he become committed to his own conditions for peace. Indeed, inflicting more damage on the enemy might cause him to stiffen his peace terms: in World War I the destruction wrought by the Germans in Belgium and northern France contributed to the French and British stance against a compromise peace with Germany.

To put it differently, the conditions on which both sides can agree for ending the fighting are not independent of the level of fighting. Hence, escalation that falls short of defeating the enemy may cut both ways. On the one hand, it may induce one side (or even both) to seek new ways for ending the war since the costs and risks of fighting have become harder to b∘ar. On the other hand, it may raise the ambitions on one or both sides and thus widen the gap between what one side would settle for and what the other demands. It is these opposed effects of escalation that make it so hard to plan for limited wars and to terminate them.

PITFALLS OF ANALYSIS

Germany's unlimited submarine campaign in World War I is an example of an escalation that was meant to end the fighting quickly. The decision by the German government to undertake this campaign was preceded by prolonged debate at the highest level. Finally, the conclusions of a comprehensive analytical study submitted to the high command provided crucial support to those who wanted to go ahead with the campaign. The secret records of this debate and the supporting policy research were made public after World War I, and additional documents from German archives became available after World War II. We can trace here the use of something akin to a modern systems analysis and its impact on government decisions.

Yet, egregious mistakes were committed in this analysis— mistakes that lost an empire. They exemplify the pitfalls of policy studies done by advocates. For this reason, the 1916 German precursor of a military systems analysis is highly instructive today.

The German analysis predicted that the submarine campaign would force England to make peace within five months, thus promising a short road to victory. This prediction was crucial for the final decision to go ahead with escalation.[2] Within the German government, the analysis recommending unlimited submarine warfare was put forward, of course, by the naval staff. It was submitted in a lengthy memorandum to Hindenburg—the highest military authority—on December 22, 1916. Shortly thereafter, on January 8, 1917, the decision to launch the unlimited submarine campaign was rendered final by the Kaiser.[3]

At first sight, the memorandum prepared by the German navy gives the impression of having carefully weighed all the relevant aspects of the question. It brings together a broad array of statistics about England's supply of essential raw materials, discussing wheat and coal, cotton and wool, oil and timber. The analysis seems to lean over backwards to estimate conservatively the tonnage that the proposed submarine campaign could sink. It makes careful allowance for various ways in which the British might compensate for their lost shipping, such as acquiring the German ships that were blocked in American ports, or reducing shipping needs by withdrawing British troops from the front in Greece.[4]

The basic technical estimate in this analysis regarding the physical effect of the proposed escalation was properly conservative. The naval staff estimated that the new campaign would sink 600,000 tons a month; the average during the crucial five-month period (within which England was predicted to sue for peace) turned out to be 658,000 tons.[5] The analysis was also not far from the mark regarding the effects of this loss in shipping on the supply of foodstuffs in England. In April and May 1917—two months after the campaign had started—supplies of wheat, flour, and rice were lower than had been estimated in the navy's analysis. By the end of April, the past year's potato crop was practically exhausted, and on April 25, 1917, the British Minister of Food announced that the food supplies might not last until the next harvest without further economies in bread consumption.[6] Though the British never instituted food rationing, the discrepancy between supply and demand led to queueing and large price increases.

As required for any good cost-benefit analysis, the German naval staff also considered political and strategic side-effects of the recommended campaign. The memorandum admitted the possibility that some of the neutral European countries, such as Holland and Denmark, might join the Allies in the war against Germany as a result of the unrestricted submarine warfare. Indeed, during the preceding year, German leaders who still opposed this escalation stressed the risk of Holland's entry into the war. And as late as September, 1916, Ludendorff had remained opposed to the submarine campaign for fear that it would drive Denmark into the war against Germany, at a time when the situation of his army seemed critical.[7]

A few months later, the German army had overcome its temporary weakness and the military threat from a possible intervention by little Holland or Denmark seemed a tolerable risk to take. However, this still left the possibility that the United States—the strongest of all the neutral powers—might enter the war against Germany as a result of the submarine campaign. The analysis of the German naval staff bravely confronted this contingency, too.

With our hindsight, we know, of course, that the American entry into the war was the decisive factor in the ultimate defeat of Imperial Germany. How did the German analysis cope with this contingency? It started out on the right track: "It is appropriate (*zweckmässig*) to consider the less favorable outcome as probable and to clarify what effect it would have on the course of the war if America did join our enemy. As far as shipping capacity is concerned, this effect could only be small. . . . One should attribute just as little effect to American troops, which could not be transported in large numbers anyhow for lack of shipping, and to American money, which cannot replace missing transport capacity."

Then followed the most crucial mistake of the entire analysis. The evaluation of the effects of America's entry was based exclusively on the prediction that England would sue for peace within five months. From here on, everything else fell into place nicely. No need to worry about the later American contribution to the war: "There remains only the question, what America will do in view of the peace settlement that will be forced upon England. It is unlikely that America would then decide to continue the war against us alone, since it would be without the means to attack us forcefully." [8] Indeed, given the crucial, but mistaken, prediction, this conclusion seems most plausible and would appear reasonable still today.

During the debate in 1916, the German Secretary of State assured those among his government colleagues who still opposed the escalation that the American contribution to the war would be nil. As he sheepishly explained in the postwar hearings, he had been referring in 1916 only to America's contribution within *these five months:* "The possibility that the war could still last two years despite the unlimited U-boat warfare," he revealed, "was not being considered seriously by anyone at that time." Even after America had entered the war on April 6, and the unlimited submarine campaign had already lasted for three months, Ludendorff remained convinced that he need not worry about the American mobilization potential. Germany's ambassador to the United States, upon returning from Washington, warned Ludendorff how strong an army the United States could mobilize and send to France. Ludendorff replied that Germany would have plenty of time to terminate the war (with the United States) before that, because within three months the submarine campaign would force England to make peace. [9]

Thus, mistake number one in this analysis was to treat a prediction (England's suing for peace within five months) as if it were a certainty, instead of directing attention to the possibility that the prediction might be wrong. Indeed, this prediction was slipped in implicitly, as if it formed a most unchallengeable assumption. Mistakes of this type can also occur in contemporary government studies. They are facilitated by introducing such a central prediction by implication, instead of stating it overtly, recognizing its decisive role, and giving reasons for assigning an overwhelmingly high probability to it.

Even though the advocates estimated that chances of success were high enough to warrant going ahead with the campaign, they should have recognized the desirability of keeping the United States neutral—as an insurance measure, so to speak, in the (to them) unlikely event that England should fight on for *more* than five months. Instead, the Germans were highly cavalier about provoking President Wilson into a declaration of war.

In modern jargon the submarine campaign was meant to be a *coercive* measure, whose effects in interdicting the enemy's military supplies were secondary. That is, the proponents of the campaign never advocated it as a means to cut the enemy supplies to the battlefields in France; they advocated it as the means to make England decide to quit the war on Germany's terms.

Given this view, one would expect the German advocates of the submarine warfare to have taken great pains in estimating the calculations and decisions of the British government, and to have recommended that the pressures of the blockade be coupled with political incentives. Everything in their plan hinged on the predicted decision in London; hence all efforts should have been bent to make that prediction come true.

Nothing of this sort was done. No political inducements were offered to the British to render it more likely that they would, in fact, make peace. And while the German advocates of the submarine campaign conscientiously included the contingency of America's entry in their analysis, they failed to reexamine their prediction about England's suing for peace in the light of it. They borrowed this prediction from the first part of their study, where America's entry had not yet been considered, and simply inserted it into the second part of the study, which dealt with the effects of America's entry. That is, they argued as if England's decision to sue for peace were independent of America's joining the war on her side. Incidentally, the opponents of the campaign—while its pros and cons were still being debated within the German government—failed to drive this capital point home, although many German leaders did question the assumption that England would sue for peace. (The Kaiser himself said in March, 1916, that if Germany threatened to defeat England in naval warfare, every Englishman would fight to the bitter end before capitulating.) [10]

How did the advocates of escalation elaborate the key assumption that England would give in within five months; what data did they adduce, and what was their reasoning? They jumped from the relatively accurate predictions of shipping losses and food shortages to the political forecast regarding the reactions of the British government. This was done by using metaphorical language—a common source of error in political analyses: "If it is possible to break England's back, the war will be decided in our favor immediately. But England's backbone is her shipping . . ."

Once this point about Britain's vertebrate anatomy was settled, one could elaborate questions of psychology. The British should be left in no doubt regarding the seriousness of the German intentions, the analysis argued. Immediately after the campaign of unlimited submarine warfare was announced, it should be put into effect; no intervening diplomatic discussions should be permitted to give the British false hopes. "If this is done, then, but only then, will the kind of panic seize the shipping circles, the English people, and the neutrals, that guarantees the success of unlimited submarine warfare. This success is to be expected with certainty within a period of five months at the most. The success will suffice to make England inclined toward an acceptable peace." [11] A jumble of metaphors and fuzzy language: England's backbone would be broken, panic would seize her shipping circles and her people, and England would become "inclined" toward an "acceptable" peace.

The internal German debate about the submarine escalation provides illustrations of other classical mistakes that can occur in this kind of analysis if done badly, today as well as in 1916. One such mistake is to exaggerate the difference between the effects of existing policies and the effects of the new, recommended policy. A more restricted kind of submarine warfare (that minimized the risk of provoking America's entry into the war) was already underway at that time. The German naval staff estimated that this restricted warfare would have destroyed 18 percent of British shipping within the axiomatic five months, as contrasted with the estimated 39 percent to be sunk by the recommended unlimited campaign. The German analysts were as certain that the 18 percent sunk would not suffice, as they were certain that the escalated level would bring full success: "I consider it as impossible that England, under Lloyd George's leadership with his extreme determination, could be forced to make peace [by the sinking of 18 percent of her shipping]." [12] But reduce England's residual tonnage from 82 percent down to 61 percent, with the same Lloyd George and his same determination, and the impossibility turns into certainty! One suspects this kind of reasoning about escalation has not gone entirely out of fashion during the last fifty years.

Britain's determination may in part have been due to the very weapon that the Germans used. In the autumn of 1916, before the Germans had started their *unlimited* submarine warfare, the British War Cabinet debated England's war aims and the question of a compromise peace. In a memorandum for the War Cabinet, Lord Robert Cecil argued that England should not make any peace moves, because a compromise settlement "would be known by the Germans to have been forced upon us by their submarines, and . . . our insular position would be recognized as increasing instead of diminishing our vulnerability. No one can contemplate our future after ten years of such conditions without profound misgiving." [13]

Another and common mistake is to launch upon a dangerous course of action because one cannot think of any better means to pursue one's old ends, but fail to examine whether one's ends ought to be changed. The German submarine analysis, at one point, argued that the break with America must be accepted "because we have no other choice." [14] And throughout the whole debate for and against this escalation, the advocates argued that the step had to be taken because it was the only effective means still left for fighting on.

The final lesson of this story is that those whose egregious blunders destroyed Germany's chances for a peace settlement on more favorable terms refused to learn the lesson at all. In the postwar hearings, the witness representing the former Imperial Navy, an Admiral Koch, argued that the case for escalation looked as good in 1919 as it had looked in the navy's analysis in 1916. In 1917, he said, the British would indeed have been close to seeking a compromise peace, had it not been for the peace resolution that the Socialists initiated in the German parliament and had the British not discovered that the Austrians wanted to make peace so badly. [15] That is, according to the Admiral (and many others who thought like him), the unlimited submarine campaign would have brought victory had it not been for peacemongers and leftists back home. This view soon gained currency as the argument (or legend) of "the stab in the back."

This argument is based on a purely imaginary transposition of strength from the enemy's side to one's own. It happens often after a war has ended in failure that those who advocated escalation or prolongation of fighting argue all would have ended well had only their own nation shown more fortitude but the enemy nation less. That is to say, these advocates allocate some imaginary tolerance for casualties and other deprivations to their own country, while denying the enemy—in their calculations—the degree of tolerance that he in fact has shown. It is as if a general, in his imagination, allocated some of the enemy's divisions to himself and then blamed his defeat, not on the actual balance of forces, but on some sinister machinations among his troops.

If a statesman or general commits a mistake that proves disastrous for his nation, he will require great moral strength to avow his error. Where that strength is lacking, blaming "the stab in the back" often provides the convenient self-justification that moral cowardice demands.

THE THREAT IS BETTER THAN ITS EXECUTION

We turn now to a different case, where peace was obtained through a threat of escalation rather than through actual escalation.

At the beginning of World War II, in the wake of the Nazi-Soviet Pact, Stalin tried to force Finland to cede vast territories north of Leningrad, perhaps to protect that city against the contingency of German aggression, perhaps also in order eventually to subjugate Finland and incorporate her into the Soviet Union, just as he was to subjugate the Baltic states. Since the Finnish government staunchly refused to give in, Stalin sought to take by force what he could not obtain through diplomatic pressure. This action led to the Finnish Winter War. After some five months of fighting, during which the small Finnish army defended itself heroically, Finland's defenses at last began to weaken before the massive preponderance of the Soviet army.

How could this war end, except in the total defeat of Finland? Sweden, though sympathetic toward Finland, feared to become involved; and Hitler, who had just divided Poland with Stalin, was not yet ready to attack Russia. In the West, this period was still that of "the phony war," when France and Great Britain had not the strength to launch an offensive against Nazi Germany, and Hitler's armies and air force were not yet ready for their big offensive.

Yet, curiously, the military weakness of the Western Powers did not prevent them from hatching a most bizarre scheme for coming to the aid of Finland. The French Premier, Edouard Daladier, frustrated by France's inability to attack Germany on the Western front and hostile toward the Soviet Union because of the Nazi-Soviet Pact, promoted a plan for sending a French-British expeditionary force to Finland through Norway. The British government went along with the plan most reluctantly, and even the Finnish leaders, badly in need of aid as they were, remained doubtful. They feared the proffered help would be halted at the coast of Norway, and even if it did get through Norway and Sweden (neither had given their consent!) the intervention would carry Finland into a global war without assuring it of any advantage.[16]

The essential question regarding a *threat* of escalation, however, is always how the threat looks to the other side. For Stalin, the risk of becoming involved in a war with the Western Powers while he was trying to preserve his uneasy pact with Hitler seemed precisely what he had to avoid. The French-British plans for sending an expeditionary force to Finland were not kept secret from him. Militarily unrealistic though they were, they suggested just enough risk of a new "Munich"—collusion of the Western Powers with Germany at the Soviet Union's expense—the worst possible development for Stalin.[17]

Although one cannot be certain, it is probably this felt risk that caused Stalin to refrain from occupying all of Finland, which he could have done within a few months, given his overwhelming military superiority. Instead he offered peace terms that at least permitted Finland to survive as an independent nation. (That Stalin wanted to subjugate all of Finland became clear half a year later, in the fall of 1940, when he leveled highly threatening demands on the Finns. Franco-British help was out of the question then, after the fall of France, but instead Hitler provided the necessary deterrent by warning Stalin unmistakably not to attack Finland again.) [18]

World history hung in the balance in the early days of March, 1940. As Finnish negotiators in Moscow were unsuccessfully trying to soften Stalin's peace terms, the Finnish front was nearly crumbling under the Russian onslaught. At the same time, the government leaders in Helsinki were debating whether or not to accept the unrealistic Franco-British offer for help. Only after considerable internal disagreement had been overcome did the Finnish government accept the Soviet peace terms instead. In fact, the Finnish President was disposed to continue fighting if the cabinet agreed unanimously. One cabinet member argued that it was dangerous to the nation's morale to give up before the army was beaten.[19] (Incidentally, this idea, often encountered during the crisis of war termination, is usually fallacious, for giving up *after* the army has been beaten is almost invariably worse for the nation.) Had the Finnish government decided against peace and concurred in the French-British expedition, the Western Powers (unless they had reneged on their promise) would have become involved in a war with the Soviet Union, instead of concentrating their meager strength on fighting Nazi Germany. Thus, in those days, a few votes in the Finnish cabinet could have changed world history. But as it turned out, the *threat* of escalation sufficed to end the war.

The gamble of seeking to end a war by opening a new front or through some other form of escalation has often tempted government leaders. Frustrated by a deadlock on the main front and unable to reach peace through negotiation, they may sometimes feel that the only escape left is to expand the war. One of the gravest miscalculations of this kind was committed by Adolf Hitler in 1940, when he decided to attack the Soviet Union without first defeating England.

After Hitler had defeated France, he was interested in a settlement with England. According to several indications, his peace terms initially would have been fairly mild (mild, that is, relative to Hitler's usual style), envisaging an arrangement for global spheres-of-interest. At that time, Russia posed no imminent threat to him. Indeed, Stalin was so afraid of Hitler's armies that he kept supplying him with raw materials badly needed by Germany. The very possibility of a German attack seemed to keep Stalin compliant.

From Hitler's point of view, maintaining the invasion of Russia as a threat would have been far better than its execution. Nonetheless, soon after Hitler's short-lived preparations to invade England in the summer of 1940 failed, he began to prepare the attack on Russia. The early stages in his decision to turn the might of his armies against the East were apparently influenced by his desire to force England to make peace. Once he had ensured a safe and dominant position for Germany in the East, Hitler seems to have reasoned, England would be deprived of her only hope for defeating Germany—an alliance with Russia. In a memorandum on August 22, 1941, he wrote of his war against Russia: "The aim of this campaign is to eliminate Russia definitely as a continental power allied to Great Britain and thereby to deprive England of all hope to change its fate with the help of the last major power left." [20]

Hitler desired to destroy Soviet Russia, seen as his main ideological enemy, in any event. But his timing of the attack in the East was part and parcel of his scheme for defeating England. In order to enable Germany to concentrate all her forces against England, which Hitler considered his strongest opponent, Germany first had to remove the secondary threat against it in the East. He feared, above all, a two-front war—haunted by the memory of Germany's difficulties in World War I—and he seems to have expected to eliminate this specter once and for all through a quick victory over his "inferior" Russian enemy, before attempting again to coerce Britain, or to conquer her. As in the Greek tragedy of Oedipus, the very scheme designed to avert the dreaded event turned precisely into the course of action that brought it about.

In 1950–51 the setbacks in the Korean war confronted the American government with difficult decisions about escalation. General Douglas MacArthur requested permission for air attacks on Chinese territory north of the Yalu, to beat back the invading Chinese forces. In the Congressional hearings after his dismissal he contended: "Had I been permitted to use my air, when those Chinese forces came in there, I haven't the faintest doubt we would have thrown them back." By implication, if not explicitly, MacArthur's calculation seems to have been that such escalation by the UN side would not only have "thrown back" the Chinese forces but would also have led to a favorable termination of the war within costs acceptable to the United States.

The Truman administration, however, disagreed. General Omar Bradley, as chairman of the Joint Chiefs of Staff, made clear the concern that escalation against China would fail to bring peace:

There are also those who deplore the present military situation in Korea and urge us to engage Red China in a larger war to solve this problem. Taking on Red China is not a decisive move, does not guarantee the end of the war in Korea, and may not bring China to her knees. We have only to look back to the five long years when the Japanese, one of the greatest military powers of that time, moved into China and had almost full control of a large part of China, and yet were never able to conclude that war successfully. I would say that from past history one would only jump from a smaller conflict to a larger deadlock at greater expense. My own feeling is to avoid such an engagement if possible because victory in Korea would not be assured and victory over Red China would be many years away. We believe that every effort should be made to settle the present conflict without extending it outside Korea. If this proves to be impossible, then other measures may have to be taken.[21]

The effort to settle the conflict without engaging China still took two years and cost the United States some 20,000 fatalities, almost two-thirds of the toll of the entire war. From hindsight it seems plausible that a war with China across the Manchurian border would have lasted far longer and cost much more. One can, of course, never know whether in the end the UN side would have succeeded in uniting Korea by force and in coercing China to accept such a peace; or whether the war with China would have dragged on indecisively until the American and other UN forces withdrew to South Korea, and, indeed, whether South Korea could then have been held; or whether losses and frustration would have led the United States to employ nuclear weaponry.

PLUNGE OF DESPERATION

The notion that a war might be ended in a victory by expanding it is, of course, not always wrong. Escalation can succeed, if it either helps to crush the enemy's forces, or if it brings about a change in the enemy's government favorable to a settlement. The latter is essentially what happened as a result of the atomic bombings of Hiroshima and Nagasaki in 1945—whether this result was obtainable by other means or not. The sudden vast disaster in these two cities, combined with the Soviet entry into the war, broke the political power of the die-hard faction in Tokyo. (And, as has since been told, this double onslaught was just barely sufficient; in part because the peace terms the Allies offered seemed so harsh to the Japanese military.)

By and large, when escalation—or the threat of it—has succeeded in reversing the enemy's determination to fight on, it has consisted of an extraordinarily powerful move. For instance, by confronting the enemy with a sudden doubling of the forces fighting him, escalation has some times had a decisive impact in the enemy's capital. The impact was decisive because it either helped the enemy's peace faction to dislodge leaders who were committed to fight on, or it caused a sudden change of mind in the enemy leadership. Ordinarily, a gradual buildup of forces or increased attacks that the enemy can readily absorb will not accomplish the desired effect. Such weaker forms of escalation might suffice, however, to change the military outcome and thus indirectly, and over time, through a stalemate or victory on the battlefield, make the enemy ready to come to terms.

The notion that escalation might help to end a difficult war is so deeply ingrained that it sometimes takes hold of government leaders whose country has been nearly defeated. A leader's sense of reality seems to shrink as he is being pushed closer to the agonizing choice between surrender, on the one hand, and seeing his country completely occupied or destroyed, on the other. Seeking a mental escape from this painful dilemma, he may indulge in the fantasy of some "bold move" that would miraculously change the fortunes of war.

Early in 1943, Mussolini finally began to sense that his Fascist dictatorship might be nearing the end: Hitler had lost the battle of Stalingrad, the American and British forces had landed in North Africa, and the Italian army and navy had suffered a long series of defeats. But as if to seek relief from this ominous reality Mussolini became captivated by the idea that the Axis forces should escalate their war effort in the Mediterranean. Mussolini's "plan"—if such it can be called—was that Italian and German forces should move across then neutral Spain and into Morocco. Apparently without any military calculations to support it, he put this idea to Hitler in his letter of March 26, 1943: "The day when the first German armored unit arrives in the rear of Gibraltar, the English fleet must move out. . . . Even without conquering the Rock of Gibraltar we would have—with long-range artillery—control of the Straits. . . . Cut off from supplies, the fate of the Anglo-American troops would be sealed. What I propose to you is a bold move, but you have given too many proofs of your audacity for this not to interest you." [22]

At the time of their imminent demise, not only "mad dictators" but also leaders of democratic governments have deluded themselves by such escalation fantasies. Such a plunge of desperation was planned by the French government prior to the fall of France in World War II. It grew out of the French-British plan for sending troops through Norway to Finland to help in the Winter War against Russia. When the Finnish-Soviet peace settlement removed the justification of helping Finland, the British lost interest in this plan; but for the French leaders it became the great "bold" move that would transform their dismal military reality. On the very day that the Finnish capitulation became known, Premier Daladier was instructing his ambassador in London to urge the merits of the "Norwegian operation" upon the British government: "A bold and immediate initiative on our part in Norway is necessary today to reverse [our declining prestige and] the general degradation of our military and diplomatic position, as well as our moral authority." [23]

On March 25, 1940, Paul Reynaud, who had just succeeded Daladier as Premier, went even further. He asked his ambassador in London to enlist Britain's participation not simply in the operation in Norway but also in "an equally decisive operation" in the Black Sea region, "not only to impede the German army supplies, but above all to paralyze the economy of the USSR before Germany succeeds in mobilizing it for its own profit. . . . The absence of a state of war between the Allies and Russia is perhaps being conceded by the British Government as a juridical obstacle to this enterprise. . . . The French Government is ready, if the British Government judges it necessary for military action in the Caucasus, immediately to examine . . . the best justification [for ending our diplomatic relations with the USSR]." [24]

To put it bluntly, the French Premier, sensing that his country could not bring effective force to bear upon the Germans, found relief from the oppressive reality by proposing that France and Britain should also make war against Russia.

The fantastic proposal was based on plans to bomb Baku

and other Caucasian oil fields, drawn up by French and British air force officers. Apparently, the question whether or not this action might provoke Soviet entry into the war on the side of Germany against France and England, and if so, how this would affect the outcome, was never responsibly considered. (In this, as well as in other respects, the plan was inferior to the analysis by the German navy which recommended the escalation of the submarine warfare in World War I. The latter at least tried to consider the effects of American entry into the war.) The commander of the French army, General Gamelin, had concluded a month earlier that the proposed action against the Caucasian oil fields was "of great interest. It would permit a very serious, if not decisive, strike at the Soviet military and economic system." [25]

Such was belief in the effectiveness of airpower that some 177 planes were supposed to destroy 75 percent of Soviet oil production. The French version of the plan—which, incidentally, neglected to take account of probable aircraft losses—promised that the paralysis of the oil industry "would have incalculable consequences." (That famous phrase, "incalculable consequences," so often seems to cap recommendations for "bold moves.") The British version forecast that the achievement of the 177 planes would induce "sooner or later the total collapse of the war potential of the USSR." It promised that the recommended operation could "decide the course of the entire war." [26] Indeed, so it could have, by cementing the Nazi-Soviet Pact long enough to ensure the defeat of England.

FOUR

THE STRUGGLE WITHIN:
PATRIOTS AGAINST "TRAITORS"

> *He who dares to act must do so in the knowledge that he will go down in German history as a traitor. But if he failed to act, he would be a traitor to his own conscience.*
> —CLAUS GRAF STAUFFENBERG, *before he attempted to assassinate Hitler, July, 1944*

NOTHING IS MORE DIVISIVE for a government than having to make peace at the price of major concessions. The process of ending a war almost inevitably evokes an intense internal struggle if it means abandoning an ally or giving up popularly accepted objectives. Concessions required for peace, however, are not the only cause of internal disagreement. When the fighting comes to an end, the heavy toll that the war has taken—like a debt that comes due—may suddenly contribute to dissension at home. Indeed, after prolonged and costly fighting, not only the losing nations but also the victors are often torn by political upheavals.

The power structure of a govenment is not made of one piece —even in dictatorships. Political factions contend for influence, government agencies and military services maintain their own sep-

arate loyalties and pursue partisan objectives, and the basis of
popular support keeps shifting. During a war, different parts of
this power structure become differently committed to the military
effort. The more important a group's role in this effort, the greater
the share of the nation's resources on which it can exert a claim.
The war's end will undermine this newly gained influence and may
bring other factions to the top.

Those to whom a war has brought political power or economic benefit will nevertheless accept an end to the fighting without opposition—if the enemy surrenders unconditionally. World
War II saw such a conclusion. But when the attainable peace
terms fail to satisfy earlier war aims, powerful men and their
supporters may—consciously or unconsciously—try to maintain
their private advantages and political positions by objecting to the
disappointing settlement.

WHO ARE THE TRAITORS—
THE DOVES OR THE HAWKS?

In the internal struggle about ending a war, each faction invariably argues that it wants "peace with honor." Yet a prolonged
war can bring such deep disagreement on national objectives that
this phrase has no common meaning. In contrast to the idea of
"peace with honor" are set the various concepts of "betrayal"—
betraying an ally, betraying the sacrifices that have been made on
the battlefield, betraying the nation's ideals. Those who wish to
end a war risk exposing themselves to charges that they are promoting such betrayals, unless, of course, all the factions in a country are agreed that the advocated peace terms would mean victory.

Close behind charges of betrayal lurks the accusation of
"treason." The threat of being charged with treason is a powerful
deterrent to those who want to promote a peace settlement with
the enemy. The importance of the threat, though, lies not in legal
sanctions but in moral condemnation. Fear of this taint of "trai-
tor" deters senior officers and government officials from taking

steps to end a war, even if they know full well that further fighting will do more harm than good.

There are valid reasons why treason should evoke such violent and condemnation and be subject to the highest penalties. The horror of "this most odious crime" helps to keep a nation, once committed to a war, from breaking up in internal disagreements. Also, a nation at war is acutely vulnerable to disloyal acts by military men or leading civilian officials who choose "to give aid and comfort" to the enemy. Throughout the ages, states have sought to protect themselves against this threat by strong moral and legal sanctions. Defenses are much weaker, however, against internal threats to the survival of a nation that stem from obstinacy in fighting on for unattainable aims or from starting a war through wanton acts. A dangerous asymmetry exists here in the protection of a state against two types of harmful acts by its own citizens.

The English language is without a word of equally strong opprobrium to designate acts that can lead to the destruction of one's government and one's country, not by giving aid and comfort to the enemy, but by making enemies; not by fighting too little, but by fighting too much or too long. "Adventurism"—much too weak a word—is perhaps the best term to describe this "treason of the hawks." (The term "war crimes" does not have the same meaning. Charges of "war crimes" in past practice concerned either atrocities committed during combat or military occupation, or the guilt, as an offense against the attacked country, of initiating a war; they did not concern the guilt of initiating or prolonging a war seen as an offense against one's own country.) [1]

Treason can help our enemies destroy our country by making them stronger; adventurism can destroy our country by making our enemies more numerous. Treason can bring us defeat by retreating in the face of the enemy; adventurism can bring us defeat by advancing till our forces are overwhelmed on distant battlefields. Treason can force us into capitulation by treating secretly with the enemy; adventurism can force us into capitulation by failing to treat soon enough with the enemy. Treason can enable our enemy to break our alliances apart; adventurism can enable our allies to pull us down into disaster. It is hard to say whether treason or adventurism has brought more nations to the graveyard of history. The record is muddied, because when adventurists have destroyed a nation they usually blamed "traitors" for the calamity.

The horror of "treason" sometimes goes so far that it leads to strangely perverted loyalties. In the final stages of several wars, senior government leaders knew beyond any shadow of doubt that by remaining loyal to an ally their country would meet with total disaster. Yet they would rather betray the interests of their country, perhaps even in violation of orders from the head of state, than betray their ally in violation of a treaty.

The demise of the Hapsburg monarchy in World War I might be attributed to such a perverted sense of loyalty. The Austrian leaders could see, long before the final collapse in the autumn of 1918, that the war would be lost; and they also realized that Imperial Austria would not survive as a state unless peace were made quickly. In April, 1917, the Austrian Foreign Minister, Czernin, prepared a memorandum for his Emperor to be sent to the German Emperor. In it he argued that by summer or the following fall, Austria and Germany should terminate the war "at any price." The reason he stressed as principal was the danger of revolution at home: "I am convinced . . . if Germany should attempt to conduct another winter campaign, that there will be basic changes within the Reich, which to me, as the responsible defender of the dynastic principle, seem much worse than a bad peace concluded by the Monarchs [i.e., the Austrian and German Emperors]." [2]

Despite his perfect foresight, the Austrian Foreign Minister failed to be a "responsible defender of the dynastic principle." He betrayed the wishes and interests of his own Emperor to avoid "treason" against the German ally. He knew that his Emperor Karl was anxious to make peace. He knew the military weaknesses of Germany and Austria. He realized that the unlimited submarine campaign, which the Germans had initiated against the insistent pleadings and better judgment of the Austrians, had turned out to be a disaster. And he was fully aware of the longing of the Austrian people for peace, and the threat of total disintegration of the Austrian Empire. All the same, he would have nothing to do with a separate peace.

He could have explored several promising opportunities. For instance, not long before he drafted his prophetic memorandum, he was approached by the American ambassador in Vienna with the suggestion that the Austrian government could, by "indicating a desire for an early peace," obtain assurances from the Allied Powers that they did not wish to separate Hungary and Bohemia from Austria. Czernin's reply was that Austria "could only enter into negotiations for peace simultaneously with her allies." American Secretary of State Lansing urged Czernin to reconsider and warned him that he should secure for his country certain advantages which might not be obtainable for long. To no avail. On March 13, 1917, Czernin told the American ambassador that it was "absolutely out of the question to separate Austria-Hungary from her allies." [3] Yet Austria had no quarrel with France or England, much less with the United States.

Even in September, 1918, when Austria's primary enemy, Russia, had long been eliminated, and when the failure of the German strategy against the West had been admitted even by the German High Command, the Austrians still felt qualms about deserting their ally. Late in August and through September, the Germans successfully dissuaded the Austrians from their intention to make an immediate request for peace. The Austrian Emperor told the German representative, who importuned him to abstain from his own peace initiative, that German doubt of his "fidelity" to his ally was "actually an insult to him." He took for granted that obligations to a foreign power took precedence over his obligation to defend the interests of his own country and crown. At that very

time, while inhibiting the Austrians with the imputation of treason, the Germans still refused to inform them of Germany's war
aims.[4]

The misplaced loyalty of the Austrian government leaders
cannot be explained by the fear that the Germans would have intervened by force had the Austrians made a separate peace. The
possibility of German intervention so late in the war, and the dis-
advantages that might result from it, were outweighed by the prospect that a continuation of the war would lead to the dissolution
of the state—a prospect about which the Austrian leaders were in
no doubt. To be sure, in November, 1917, while weighing the possibility of a separate peace in a letter to a friend, Czernin did
voice his fear that Germany would occupy Austria: "Instead of
bringing the war to an end, we would be merely changing one opponent for another and delivering up provinces hitherto spared.",
However, he equivocated and argued that *after* a new German
success (supposed to become possible because of the imminent
peace with Russia) he would propose to make peace "even at a
loss." [5] Would not a newly won battle have given Germany even
more power to intervene against a separate Austrian peace?

Arguing that the time for peace through concessions will
come only after a military success is achieved is a common form
of procrastination when the time has come to end a lost war. What
one of the leading British "doves" in World War I, Lord Lansdowne, had to say in March, 1918, still seems valid today:

We shall be told that the moment when the Allied Armies are achiev-
ing glorious successes in the field is not the moment for even hinting
at the possibility of peace: if the hint had been thrown out at the
moment when the fortunes of war were turning against us, we
should have been told still more emphatically that that moment, too,
was inopportune and that we must meet our reverses with a bolder
front.[6]

The drama of alliance loyalty takes a different ending in Finland's fight with the Soviet Union during the last years of World

War II. To make peace, Finland also faced the problem of "betraying" her German ally. And, as in the case of Imperial Austria, ending the war was not a problem of predicting its outcome; after early 1943, the Finns were as clear as the Austrians had been in the closing years of World War I that Germany and her allies would meet total defeat. Had those Finnish leaders who wanted to remain loyal to Germany prevailed, Finland as an independent nation would almost certainly have been extinguished more completely than the old Austrian Empire.

In the spring of 1943, while the Finnish military chief, Mannerheim, was in Switzerland recovering from an illness, the leaders in Helsinki almost signed a pact with Hitler prohibiting any separate peace. Mannerheim, who had no doubt about Germany's impending defeat, intervened just in time to prevent this. Then, throughout 1943 and the early part of 1944, the Soviets repeatedly discussed peace terms with the Finns. In March, 1944, while Finland still fought on the German side, a Finnish delegation even went to Moscow. But the Finnish government continued its dangerous policy of procrastination and turned down the Soviet terms. Under the pressure of a Soviet offensive in June, 1944, Finnish President Ryti became even more foolhardy. In order to obtain German arms, he sent a letter to Hitler pledging to make peace only in agreement with Germany.[7]

At this time, the Russians were poised to drive the Germans out of the Baltic states. And the United States, which had long tried to induce Finland to make a separate peace, finally broke diplomatic relations with Helsinki. Finland seemed doomed to become a Russian province again. A month later, however, a political miracle happened in Helsinki. The Finnish government managed to make a 180-degree turn without impairing the cohesion of the army or provoking a coup by the pro-German faction. Ryti, compromised by his deal with Hitler, was induced to resign, and Mannerheim, as the new President of Finland, made the necessary concessions to Russia while expelling the German troops from

Finland. (The German forces, enraged by Finland's "treason," caused great devastation before they were driven out.)

Mannerheim, the old military hero, was the man who could least be accused of treason. Precisely for this reason he was best equipped to lead his government during the "betrayal" of the German ally and the acceptance of harsh peace terms. His letter to Hitler, announcing the decision to make a separate peace with Russia, eloquently expresses his foremost concern—the survival of the Finnish nation:

I wish especially to emphasize that Germany will live on, even if fate should deny you victory in your fighting. Nobody can give such an assurance regarding Finland. If this nation of barely four million be defeated militarily, there can be no doubt that it will be driven into exile or exterminated. I cannot expose my people to such a risk.[8]

In 1940, during Finland's first war with Russia, Mannerheim had also helped to make peace by ensuring the army's support. His generals had favored fighting on, but in the nick of time he convinced them of the need for accepting the Russian terms.[9]

Not uncommonly the military hero of an earlier day is the man who can best conclude a "peace of betrayal." Whenever the price must be paid for ending an unsuccessful war, an ally or some groups at home will feel betrayed. It may take someone whose prestige as a patriot and great soldier is strongly established among the population at large to survive in the power struggle despite the inevitable charges of "treason"—survive long enough, at least, to put an end to the war.

When the Germans had defeated the French army in 1940 and could have occupied all of metropolitan France within a few days, it was Marshal Pétain, the World War I hero of Verdun, who led the French government that accepted Hitler's armistice conditions, betraying in a sense the British ally.

Pétain was brought into the cabinet by Premier Paul Reynaud to broaden the support of his government as the victorious German armies advanced into France in May, 1940.[10] Six days

later, on May 25, 1940, the French war council held a session
where the idea of a separate peace (or armistice) was already
being broached by various government officials. Lebrun, then President of the Republic, argued: "To be sure, we have signed agree-
ments [with England] which prohibit a separate peace. Nonetheless, if Germany offers us relatively advantageous conditions, we
must examine them very closely and deliberate them calmly." And
Pétain inquired whether there was complete reciprocity in obligations between France and England. "Each nation has obligations
toward the other," Pétain argued, "in the proportion of the aid
which the other has given it. England committed only 10 divisions
to the battle while 80 French divisions are fighting." [11] Nobody in
this council seemed to rule out negotiations with Hitler, as Pétain
kept applying the thin edge of the knife on the alliance with Britain.

Nonetheless, the "betrayal" of England (and of French pride)
had to be approached gingerly. A fortnight later, Pétain had prepared a paper to the effect that France must ask for an armistice,
but he was "still ashamed to hand it" to Prime Minister Reynaud
—as Reynaud put it to Churchill. [12] Within five days, Pétain did
not have to be "ashamed" any longer: Reynaud lost the courage of
his convictions and chose to turn the government over to Pétain,
so as to let him be responsible for the armistice.

Churchill, of course, did not consent to Pétain's "betrayal."
In one of the last British-French discussions on June 11, 1940, he
delivered a long speech advocating continued resistance against
the rapidly advancing German armies. He recommended that the
French fight in Paris, describing how a great city, if stubbornly defended, could absorb immense armies. Pétain disagreed, saying:
"To make Paris into a city of ruins will not affect the issue." [13]
Today, from the distance of hindsight, it seems that Pétain was
right on this point—the destruction of Paris would not have affected the outcome of the war.

The winner must be careful not to press too far or too fast if he wants the loser to make a separate peace. Otherwise he may upset the uneasy political truce in the losing country that keeps the government together while the "doves" overrule the "hawks."

Hitler apparently used just the right amount of moderation to accomplish his war aims through an armistice with the Pétain government. He wanted to separate France from England and to prevent the French government from leaving for North Africa with part of the fleet. To achieve these ends he laid down four conditions: (1), Pétain should not be undermined by being asked for more than his supporters would tolerate; (2), part of metropolitan France must remain unoccupied (Hitler's military planners had unimaginatively proposed occupation of *all* of France); (3), no demands should be made regarding the French colonial empire (such demands, Hitler reasoned, would merely cause the colonies to realign themselves with England); and (4), Mussolini was to conclude his own separate armistice with France, once the French-German armistice had become an accomplished fact.

Hitler induced Mussolini to abandon his plans for the occupation of the Rhone Valley and parts of French Africa. Mussolini had become Hitler's ally in the war against France only at the last minute, and his military offensive turned out to be a failure; the Italians were almost routed by the French. Thus the triumphant Hitler had no difficulty whatsoever in "betraying" the interests of his Italian ally for the sake of the armistice with Pétain.[14]

Ironically, General de Gaulle, the hero of World War II who had defied Pétain's "treason," was the man who in 1962 surrendered to the demands of the Algerian rebels. Like Pétain in 1940, de Gaulle came to power in 1958 because a war had undermined the stability of the existing government and created a need for a popular hero. De Gaulle's initial political support came largely from those groups who wanted to continue the war in Algeria till the rebels were crushed.

To end the war in Algeria, de Gaulle had not only to renege on his own promises but also to betray the Algerian "allies" of

France, those Algerians who had sided with the French forces. Indeed, it was the latter betrayal which was hardest for the French military to swallow, since they had given the most solemn vows to their Algerian supporters that they would never abandon them. De Gaulle did abandon them; and the new Algerian regime took revenge on Algerians who had remained loyal to France.

In a war where the enemy's forces invade the homeland, any government that tries to make peace with the enemy while facing military defeat will almost inevitably come apart at the seams. For one thing, the winning nation often insists on the removal of the men who led the war against it. But even apart from any such demand, the government on the losing side must turn its own political support upside down in order to reach the decision to sue for peace. Those who favored prolonging or escalating the war must either be expelled from the government (unless they prefer leaving it of their own accord), or admit that they have been wrong. On the other hand, the "doves" and appeasers will be vindicated, having favored concessions that might have either ended the war sooner or avoided it from the outset. Thus the losing nation's government must overcome a double crisis: it must grant the concessions that the enemy demands as the price for peace, and at the same time it must change its leadership and domestic support.

The crisis can become so acute that those who wish to continue the war and those who wish to end it will use violence against each other. But the "hawks" often have an advantage in obtaining their way by force, since they usually control more of a country's military establishment than the "doves." Or they can launch a new attack on the enemy during those delicate hours when an armistice is to begin or a capitulation to be consummated. The "doves" may have only their bare hands and mighty little time to stop such a misuse of military weapons, intended less to hurt the enemy than to kill the peace. Of course, the "hawks" do not proclaim that they are opposed to peace, but assert that they want a "peace with honor" instead of the settlement that is about to be obtained.

A crisis in ending a war that involved simultaneous insubordination by "hawkish" and "dovish" factions occurred in 1918, when Germany was about to negotiate an armistice with the Allied Powers. President Wilson, on behalf of the Allies, demanded as a precondition to the armistice that Germany cease her unrestricted submarine warfare. As soon as the Kaiser and the new German Chancellor had accepted this stipulation and issued the necessary orders, the German Naval Command decided on a big, last-ditch offensive by the surface fleet. Without informing the Kaiser or the Chancellor (who was about to face the enemy in the armistice negotiations), senior officers of the German navy, under the leadership of Admiral Scheer, developed a plan for a major attack on the British fleet. Admiral Scheer merely told his superiors that "the Surface Fleet has regained its full freedom of action" since it was no longer needed to protect the submarines; the Kaiser's consent to this fuzzy statement, he subsequently claimed, had authorized him to launch the new offensive.

However, within this revolt by the "hawks" there occurred a revolt of the "doves." Unrest broke out among the sailors when the ships were to be readied, preventing several of them from putting to sea. Some of the mutinous seamen made it quite clear that they wanted to protect the process of ending the war. A naval offensive now, they argued, would accomplish nothing except render armistice negotiations more difficult. The result was that the mutiny of the sailors forced the admirals to abandon their planned revolt.[15]

From the documents since released, it is possible to calculate what effect might have ensued if the revolt of the "hawks" had prevailed over the mutiny of the "doves." Had the naval offensive indeed succeeded in sinking part of the British fleet, it seems quite clear that the armistice conditions would not have been improved for Germany—contrary to the arguments later used by the German admirals in justifying their conduct. On the very day the ships were to sail, the Allied leaders met in Paris to decide on the armistice terms. The Allied commander in chief, Marshal Foch, argued against the British that the Germans should be permitted to keep their surface fleet: "What do you fear from it? During the whole war only a few of its units had ventured from their ports. The surrender of these units will be a manifestation, which will please the public, but nothing more. Why make the armistice harder, while I repeat its sole object is to place Germany hors de combat." [16] The Allied Powers adopted a compromise position; the German surface fleet was to be interned in neutral ports. It seems probable that if this last-minute attack had been carried out, the Allies would have demanded that the whole German fleet be surrendered. (As a result of later developments, part of the German fleet was not interned in a neutral port but in Scapa Flow.)

Japan's surrender in 1945 was almost thwarted by the insubordination of a fanatic military group. To inform the whole nation of the surrender and to ensure that all Japanese forces would lay down their arms, the cabinet carefully prepared a speech for the Emperor, which was recorded for broadcast the following day. But a group of young officers in the War Department tried to seize this recording and prevent its being made public. They were foiled only because a foresighted and loyal official had hidden the record. Even after the Emperor's surrender announcement had been broadcast, various contingents revolted, and demanded that the war be continued. Indeed, five days after the surrender, a coup was being planned. There was a general breakdown of loyalty and discipline within the officer corps, stimulated by a fervor for continuing the war regardless of the consequences.[17]

By contrast, in Germany in 1945 there could have been no more "hawkish" position than that of Hitler himself. He was determined to see the fighting continue to the last soldier and was willing—indeed during his last days eager—that Germany be destroyed in the process. Hence, it was the "doves" who had to resort to insubordination to end the fighting and avoid further destruction.

Accordingly, the story of ending World War II in Europe is largely a story of "treason" by the "doves." The German general in charge of the German occupation forces in Paris, Dietrich von Choltitz, disobeyed Hitler's orders to destroy the monuments of Paris before giving up the city. Nazi Minister Albert Speer, who had been in charge of war production and had been close to Hitler almost to the end, systematically defied Hitler's scorched-earth policy, circumventing and disobeying Hitler's orders to destroy Germany's industry before the advancing Allied armies. The surrender of German forces in northern Italy was facilitated by the insubordination of SS General Karl Wolff, who negotiated with the United States through Allen Dulles, the American intelligence chief in Switzerland.[18]

General Wolff's role demonstrates a paradox in the struggle between "doves" and "hawks": in a losing country, men far to the political right can sometimes work to end a war more readily than can officers and politicians in the center who feel more constrained to abide by formal rules and to observe existing obligations. Of all the men in Hitler's entourage, Heinrich Himmler, the leader of the SS and chief of the Gestapo, committed treason by negotiating secretly with the Western Allies during the final weeks of the war, attempting (unsuccessfully) to arrange a separate peace in the West. By contrast, many German generals who were ideologically less committed to Nazism or even opposed to it, and who were fully aware of the hopelessness of further fighting and the madness of Hitler's conduct of the war, nevertheless felt honorbound to remain loyal to their Führer right to the very last bizarre days of destruction.

The most concerted effort by Germans to end Hitler's war was the abortive Twentieth of July Movement. Its leaders, as early as 1939, tried to establish contact with the Western Powers and seek mutually acceptable peace terms. After these early attempts at peacemaking failed, the group realized that the Allies would not negotiate with Hitler; many of them hoped, however, that the Western Powers would deal with Germany separately at the expense of Russia. To make peace negotiations possible, they attempted to assassinate Hitler on July 20, 1944, and the failure of this attempt led to the capture and execution of most of the members of the group.

Since 1945, these actions have been widely debated in the German Federal Republic. Many argued that Hitler's criminal reign justified insubordination and treason. Senior officials of the Federal Republic repeatedly praised the Twentieth of July Movement, and the day is still commemorated there.[19] Yet, such official support for acts of treason against yesterday's head of state have not met with unanimous approval. In part, the disagreement stems merely from some leftover Nazis protesting official praise for the would-be assassins of Hitler; but, in part, it may reflect a more reputable concern as to whether treason, in principle, should ever be defended.

The condemnation of past acts of treason, though, tends to be one-sided. By and large, the self-styled patriots excoriate acts of treason only if committed by the "doves" while tolerating or even defending them if committed by the "hawks." Yet, treachery by the "hawks" can result in the gravest damage for a nation (as would have been the case had the German admirals in 1918 successfully launched an unauthorized offensive on the eve of the armistice negotiations, or had Japanese officers in 1945 thwarted the Emperor's order for surrender).

Most men would agree that the value of loyalty depends on the object of loyalty. But does the value of courage also depend on what the acts of courage are aimed at? And is the value of honor equally conditional? The bravery of German soldiers who fought for their Führer to the very last may seem meritorious if viewed in a narrow personal perspective as acts of self-sacrifice and human courage; but seen in a larger context this bravery merely prolonged one of the most criminal of regimes. The loyalty of some German generals who fought for Hitler till he killed himself in his bunker, although they were not believers in Nazism, may seem the proper conduct for an officer who has sworn allegiance to his head of state; but seen in a larger, historical perspective such loyalty was merely unthinking adherence to a professional code, which in this case meant assisting a madman in the destruction of his own nation.

Like no other crisis in society, the closing days of a destructive war force people to consider the proper precedence of values. Vices may become virtues, virtues vices. As the social order and political institutions become unhinged, everyone—from general or statesman to soldier or citizen—has to search his own conscience for guidance about the rules of ethics he should follow in seeking to serve his nation. Should compassion come before loyalty? Should reason win out over honor? Should prudence overrule pride?

DEBATING WHAT THE WAR IS ALL ABOUT

The political dissension at the time of a conditional surrender or a costly compromise peace is inevitably foreshadowed in the internal disagreements about war aims and peace negotiations, long before the decision to terminate the war is final. That is to say, the struggle between "doves" and "hawks" over when and under what terms the fighting should stop is preceded by disagreements over what the war is all about. Those officials who are identified with the initiation of the war or its early conduct are apt to fear that they would in fact be criticizing and undermining themselves as government leaders if they considered any conclusion to the war that did not achieve the principal war aims. And those who favor a compromise peace must attack—implicitly or explicitly—those who lead the nation in the war.

During World War I, war aims were debated on both sides,

particularly in Germany, Austria, and England. Parliamentary debates helped to air the disagreements in public. In the United States, however, President Wilson kept the formulation of the official war aims largely to himself, confiding to and being influenced by Colonel House, to the virtual exclusion of the Secretary of State, the military leaders, and Congress. The public discussion in the United States in World War I, by and large, favored hard terms for Germany.

In Imperial Germany, the civilian members of the cabinet did not invariably support the annexationist war aims of the military leaders, but they were cautious in pressing their disagreement.[20] In the German parliament, on the other hand, where the debate was quite open, the Social Democrats were outspoken in their criticism of those aims. To their credit, they remained critical even when the spoils could actually be collected. In March, 1918, when the German High Command forced the peace of Brest-Litovsk upon Russia, by which the Bolsheviks had to surrender a third of the population and agricultural land of the Tsarist Empire to save their regime, the Social Democrats did not join in the jubilation about the victory. Their speaker, Scheidemann, declared in the Reichstag: "We fought to defend our country from Tsarism but we are not fighting for the partition of Russia." In his opposition there was a sense of foreboding that such harsh peace terms might some day be imposed upon Germany: "We do not wish . . . to attain a dominating position which would force us to conclude a peace with the Entente on such terms as those on which Lenin and Trotsky are now concluding peace with [us]." [21]

In England, during World War I, popular support for the war effort was more unquestioned than in any other major European country. The British people pursued this most destructive war—so disastrous for themselves despite the final victory—in a jingoistic fashion that was intolerant of opposition and in stark contrast to their parliamentary traditions. They accepted war aims encumbered with imperial objectives and distant causes, while they let themselves be driven to utmost sacrifices in the foolish battles in Flanders.

It seems that the historiography of World War I (at least the British and American literature) has not focused critically upon the dismal performance of British democracy in 1914–18. Most of the earlier writings on World War I remained encrusted in the heroic patriotism of 1914 vintage. For example, a British history of the war published in 1923 records smugly that "of pacifism in the dangerous sense there were few signs, and the plea for 'peace by negotiation' sprang in most cases from an error of head rather than of heart." [22] After 1945, historians were sometimes misled by mistaken parallels to World War II. To the extent that British policy in World War I has been criticized in English writings, the targets were mostly the diplomacy prior to the war, the Versailles peace terms (attacked notably by J. M. Keynes), and particular leaders like General Douglas Haig. The malfunctioning of the British leadership and political institutions as a whole has received less attention. [23]

A few episodes from the British parliamentary debates may serve to illustrate England's political atmosphere of that time. On July 15, 1915, a member of Parliament, David Mason, asked the Prime Minister

whether H. M. Government will consider the advisability of taking steps to find out the terms of peace which the enemy Governments will entertain and whether such terms will include the evacuation of Belgium and North-Eastern France.

Prime Minister: the answer to the first part of the question is in the negative; the second part, therefore, does not arise.

Mr. S. Roberts: May I ask whether questions of this nature are not detrimental to the public interest.

On July 21, 1915, Sir W. Byles asked the Prime Minister

whether . . . he will . . . restate with more definiteness the central objects for which the nation is pouring out its blood and treasure, in the hope that some intervention of peaceful influences may lead to the attainment of these objects by other means than the continuance of warfare.

Prime Minister: I stated these objects with such definiteness as I'm capable of in my speeches at the beginning of the war.. . . .

Sir Snowden: May I ask the Prime Minister if he is aware that there is in Germany already a large and growing peace movement among the Social Democrats, and will he keep his eye upon, encourage, and take advantage of any movement for bringing this war to an early and satisfying conclusion.

Prime Minister: I have nothing to add with regard to our objects to the statements which I've already made.[24]

The policy of the British government in World War I hardly kept its eye upon, much less encouraged, "any movement for bringing this war to an early and satisfying conclusion." If one considers the constructive influence that the German Social Democrats exercised in the end in shortening the war, and how handicapped they were in their peace efforts by the lack of moderate terms or peace gestures from the Allied side, Sir Snowden's question displays far greater statemanship than the policy of the British government.

In November, 1915, Charles Trevelyan, then a member of the Liberal Party, urged in the House of Commons that England formulate her tentative peace terms, warning that a war of attrition would mean ruin for both sides:

We have started this war on the assumption that our resources are so vast that we can do anything that occurs to us, that we will respond to every Ally that moves our pity, and to every enemy challenge that excites our pride. . . . I want for a moment to ask the House to consider what a war of attrition means, or may mean. There have hardly been any wars in recent history which have been finished by wearing out one side economically, or by the sheer loss of men. . . . I see the *Times* newspaper, the chief advocate of a war of attrition, suggests that we may have to go on for ten years. Meanwhile during that war, of years, which is going to wear out Germany—I do not deny that you may be able to wear down Germany!—what is going to happen to us and the rest of the world; where are we going to be at the end of that time?

Bonar Law, Chancellor of the Exchequer and the government's leader in the House of Commons, had this answer: "A speech which could have less practical effect on any of the issues with which we are concerned never could have been made than that to which we have just listened." [25] A remarkable response in the Mother of Parliaments.

On October 11, 1916, a Labourite argued in the House of Commons:

We started the war with the noblest aims. . . . Let us be very careful that in the prosecution of this war we do not allow our objects to degenerate. . . . I hope we are, and shall be, ready to welcome anyone . . . who can come to us with a message of peace.

Reasonable as these words may sound today, to the British cabinet they seemed an outrage. Lloyd George replied for the government:

Whether or not there is going to be intervention and pourparlers to arrest the fight at the moment we are gripping the enemy is much more a military matter than it is a diplomatic matter. . . . Any intervention now would be a triumph for Germany! A military triumph! A war triumph! Intervention would have been for us a military disaster.[26]

Within the secret inner councils of the British government, at least, the 1916 debate on war aims was somewhat more thoughtful. That fall, Prime Minister Asquith asked the members of his War Committee to express their views of the military and economic prospects. Several of the memorandums that his ministers prepared predicted highly unfavorable developments. (For instance, the President of the Board of Trade anticipated "a complete breakdown in shipping . . . much sooner than June 1917." He was too pessimistic, incidentally, for despite the unforeseen intensification of German submarine warfare, there was no "complete breakdown.") But these estimates failed to budge the British statesmen from the sundry imperial war aims that their government supported, whether they were those of the Tsar, the French and Italian annexationists, or their own scheme for control of Mesopotamia.

The only dissenting voice was raised by Lord Lansdowne (then minister with portfolio). "Many of us," Lansdowne's memorandum dared to point out, "must of late have asked ourselves how this war is ever to be brought to an end. If we are told that the deliberate conclusion of the Government is that it must be fought until Germany has been beaten to the ground and sues for peace on any terms which we are pleased to accord to her, my only observation would be that we ought to know something of the data upon which this conclusion has been reached. . . . Can we afford to go on paying this same sort of price for the same sort of gains?" [27]

After this rather mild statement of dissent, both Lord Lansdowne and Prime Minister Asquith resigned; or, as some English historians calmly suggest when describing the functioning of British parliamentary democracy during World War I, this statement made the resignations necessary. A year later, although the war weariness in England had become stronger, Lord Lansdowne had difficulties in finding a newspaper willing to publish his earlier proposal that more precise and less immoderate peace terms be offered to draw out the enemy's response. The press attacked him widely for this suggestion. [28]

Fifty years later, American and British historiography has, of course, arrived at a more balanced judgment of the two sides in World War I—President Wilson's "good" allies and "bad" enemies. If it had not been for the Ludendorffs on both sides, peace might have been had in 1916. The English Ludendorffs are less revolting to our taste, for they had better manners and—until recently—were better protected by the secrecy of British archives.

In time of war, debate on war aims is opposed by the standard objection that an airing of internal disagreements would merely encourage the enemy to expect peace on *his* terms. There is some truth to this argument. As we have seen, in World War I, Germany escalated her submarine warfare in the expectation that England would sue for peace, and the voices in England opposing a relentless pursuit of the war—muffled though they were— contributed to this German delusion. On the other hand, the suppression of "defeatists" at home tends to discourage peacemakers on the other side. When the Allies brusquely rejected the German peace offer of December 16, 1916, brushing aside the more conciliatory suggestions among their own "doves," they weakened the peace movement within Germany.[29]

The long-term consequences of World War I might have been less damaging for both sides if the public and the parliaments had been allowed to participate in the formulation of war aims and to discuss strategies for ending the fighting, not just to approve budgets for continuing it. This war, that swept away or undermined the empires of both victors and vanquished, this most unnecessary war, raises starkly some fundamental questions that apply to any war: How long is it worth while to suffer—and to inflict—further casualties and destruction in order to accomplish the initial objectives of fighting? When has the time come to make concessions, so as to avoid the losses of continued warfare? Should the fighting go on to reduce the risk that the enemy will strike again in the future? Or can this very risk be better avoided by ending the war so that one's own and the enemy's population will suffer less and reconciliation might become easier? In dictatorships, short of revolt, popular feelings have little influence on these decisions. But in parliamentary democracies, too, leaders sometimes believe that the people must be protected from their erring views of the national interest—even at the price of concealing from them available choices. Who, in the last resort, is entitled to say what a war is all about?

Writing his memoirs after the war, Lloyd George discussed why it was necessary to defeat Germany so thoroughly that she would not be able to attack again, and defended himself against accusations that he had neglected to seek peace in 1916:

It is often said now by men who are seeking busily to find fault with those who shouldered the terrible responsibilities of decision in the war, that no harm would have been done had the Allies taken the initiative in approaching the Central Powers with a view to the convention of a Peace Conference in 1916. . . . . If Germany had offered to withdraw all her forces from northern France and from Belgium, merely imposing certain conditions [as to the use of the Belgian ports], could the Allied governments have aroused once more the spirit of 1914 to the pitch of facing for more than two years the horrible losses . . . merely in order to restore Alsace-Lorraine to France or to hand back Courland and other conquered territories to the incompetent hands of Russian autocracy? [30]

Is it so clear that Lloyd George's judgment was better on this choice than the judgment that he feared the British people would have made? Surely, it is not self-evident that a less stable peace— or even a less "just" peace—would have been secured had the British people decided in 1916 that it was not worth while "to fight for Courland." By continuing their sacrifices for another two years, the British people did not avoid having to "fight for Danzig" in 1939 and for their very freedom in 1940.

In the internal debate about war aims, "hawkish" factions often remain curiously blind to the fact that their own survival after the war as an influential group within the nation is endangered by the stance they assume. In this respect, "hawks" are rather "apolitical," if being "political" means to have a keen sense for the survival of one's power. In so many crises of war termination, "hawks" have grossly neglected threats to the political future of their nation in stubborn pursuit of some secondary objectives, such as territorial possessions at the periphery of their homeland or ephemeral arrangements regarding the military balance between their country and the enemy whom they happen to be fighting at that time. Engrossed by real or imagined opportunities on the military front, they redouble their efforts for short-term gains. Starting out as the defenders of the national interest, they wind up by trading their kingdom for a distant province. And in the end, when the kingdom indeed is lost, they blame their "dovish" opponents for having stabbed them in the back.

On February 5, 1918, less than a year before the empires of Germany and Austria collapsed, never to arise again, Austrian Foreign Minister Czernin got into an argument with the arch-hawk Ludendorff. Czernin managed to extract an agreement from Ludendorff in writing that their two countries were obliged solely to fight for the prewar possessions of Germany. But Ludendorff granted this concession only after vehement opposition: "If Germany makes peace without profit, then Germany has lost the war." [31] What curious inability to distinguish between the loss of some territories and the loss of the nation!

In England, too, those who insisted on pursuing World War I to victory regardless of cost hurt their own cause—the greatness of the British Empire—the most in the long run. Great Britain's position as a world power became enormously weakened. We have seen how the House of Commons shouted down, in 1916, Charles Trevelyan's warning that England's resources were not so vast that she could survive a war of attrition against Germany without irreparable damage to herself. And while the British military got their way in 1917, imposing horrendous casualties on their own country, fifteen years later the popular opposition to rearmament kept England unprepared against Hitler's aggression.

There was one government leader during the closing year of World War I, however, who understood his priorities clearly. He sacrificed a third of his country to save his regime and its long-term political objectives. Lenin—by temperament far from a "dove"—prevailed upon his Bolshevik colleagues to accept the peace terms of Brest-Litovsk. With single-minded determination, he sought to end the war at any price, lest the world's first Bolshevik government be swept away by the German armies. Germany could have advanced into Russia practically unopposed, given the disorganization of the Russian army and the overwhelming desire among the Russian people for peace. By contrast, the more "hawkish" Bolsheviks talked as if they would rather see Bolshevism disappear than lose territory (much like the "hawks" on the right): "No conscious revolutionary would agree to such dishonor," Bukharin and Radek wrote two days after the peace of Brest-Litovsk. "We should die in a fine stance, sword in hand, crying 'peace is dishonor, war is honor.' " [32]

Cutting one's losses, although a common notion in everyday life, appears to be a particularly difficult decision for a government to reach in seeking to end a prolonged and unsuccessful war. The "hawkish" and the "dovish" factions, in thinking of their country and their people, might actually not be far apart in their deeper beliefs about what must be saved. Yet, in the eyes of the "hawkish" faction, the acceptance of a partial defeat would not only expose these values to threats from without but start an internal process of political demoralization that would undermine them from within. And the "dovish" faction believes just as strongly that continued fighting would destroy these values either through some final cataclysm or through increasing strife at home.

FIVE

THE STRUGGLE WITHIN:
SEARCH FOR AN EXIT

> *It is truly unbearable for the officers and soldiers of the
> Army and Navy to surrender their arms and to face the
> occupation of the country. . . . However, compared with
> the complete disappearance of Japan, even if only a few
> seeds survive, this would allow us to envisage recovery
> and a brighter future.* —EMPEROR HIROHITO, *August 14,
> 1945, announcing to the cabinet his decision to accept
> the surrender terms*

THE POLITICAL STRUGGLE within each country af-
fects everything that matters in ending a war. It intrudes into the
formulation of the war aims, it colors and even distorts military
estimates, and it inhibits negotiation with the enemy. The views
people hold on these matters are interdependent. Those who want
their country to pursue ambitious war aims will seek out the favorable military estimates and find reasons why negotiations ought
to be avoided. Those who want negotiations to move ahead will
select the unfavorable military estimates to argue that war aims
should be scaled down.

Although at the beginning of a war, statesmen and generals

alike can remain curiously oblivious to the question of how the fighting will be brought to an end, the struggle between "doves" and "hawks" forces this question into the open. But the "hawks" can take refuge behind broad generalities, clinging to the familiar position that a "peace with honor" must be won and that the enemy will eventually have to yield since he is "hurting." The "doves" are just as frequently without a plan for peace. They may simply ask that their government start "negotiations"—as if it took only one side to reach agreement. Or, if formal talks are already taking place, they may argue that their government should offer certain concessions, and portray the settlement based on these concessions as the peace that can be obtained—as if the enemy did not have his own ideas regarding peace terms and be willing, and feel able, to fight for them.

In preparing a major military operation, military leaders and civilian officials can effectively work together in large teams to create a well-meshed, integrated plan. This holds true, almost regardless of how well or how badly the war is going. By contrast, planning to end a war where victory seems out of reach is not a task on which men can easily collaborate. To search for an exit in such a situation, government leaders can rarely move in harmony.

The sources of disagreement lie deep. For everyone involved, contradictory evidence regarding the military situation and conflicting views of the nation's priorities become intertwined with personal concerns about one's future public career, and perhaps even one's private life, after a war that had to be ended in failure.

NEGOTIATING WHILE FIGHTING

In many wars, government leaders vehemently oppose negotiating with the enemy as long as the fighting continues. They may have various reasons for this attitude: they may think an offer to talk would signal weakness to the enemy; they may be concerned that their soldiers and citizens would slacken their war efforts in the belief that peace was near; they may fear that the enemy would offer seemingly conciliatory proposals, which would be hard to reject but fall short of their war aims. Finally, they may anticipate that to raise the question of how to conduct the negotiations would stir up deep conflicts with their allies. In World War II, the agreement between the United States, Great Britain, and Russia to accept nothing short of an unconditional surrender from the Axis Powers was meant primarily to prevent a split in the uneasy alliance of the Western Powers with Russia.[1]

Whether or not negotiating while fighting will result in these effects depends entirely on circumstances. In some cases, a government has entered into negotiations with the greatest reluctance, only to discover that morale at home—if it was affected at all—rather stiffened. In other cases, the opening of negotiations caused popular disaffection with the war to erupt and bring down the government.

The more that negotiation with the enemy is presented officially as something that is natural—indeed desirable—in the midst of a war, the less will the civilian population and the troops respond to the opening of talks by questioning or rejecting a continued war effort. In the two Finnish-Soviet wars of 1940 and 1944, Finnish government leaders negotiated in Moscow while the fighting was at its peak; yet these talks did not affect the morale of the Finnish troops or the military operations initiated by either side. During the Indian-Pakistani war of 1965, the governments on both sides—having avoided a stance against negotiation—could meet for talks in Tashkent without undermining their position at home or the morale of their armies.

The very stubbornness with which government leaders sometimes oppose negotiating while fighting induces an adverse reaction at home and among the troops, once talks with the enemy have become unavoidable. In 1918, such a reaction ensued when Ludendorff, having long been the leader of the "hawkish" faction in Imperial Germany, suddenly asked his government to negotiate a cease-fire with the Allied Powers. The abruptness of this turnabout probably hastened both his own downfall (within a month he had to step down) and the collapse of the Kaiser's government. In 1917, when a large mutiny in the French army nearly caused the defeat of France, the soldiers had been demoralized not by ongoing negotiations with the enemy but by precisely the opposite: one of their chief complaints was that their government had failed to begin peace negotiations.[2]

During World War I, the principal governments on both sides stubbornly opposed negotiations while fighting, even though the ideological differences with the enemy were mild compared to World War II. There was no strong desire, in contrast to the Allies' policy toward Hitler and Mussolini, to eliminate the enemy's regime. Yet, after the outbreak of the war in 1914, diplomacy broke down completely. This stark fact has received scant attention, although much has been written about how the intricate prewar diplomacy of 1913 and 1914 might have prevented the war.

Sometimes, each side in a war refuses to negotiate for opposite reasons: one nation does not want to accept a cease-fire as a precondition to negotiation, while its enemy does not want to discuss a settlement as long as the fighting continues. This was the case in the conflict between France and the Algerian rebel organization from September, 1960, until May, 1961. In September, 1960, de Gaulle announced he would negotiate Algerian independence only if the rebels first ceased their attacks; but the rebels refused, apparently fearing that a cease-fire would lead to the disintegration of their forces or to internal strife.

Of course, such a stalemate cannot last; one or the other side will eventually yield. In May, 1961, de Gaulle yielded and began negotiations with the Algerian rebels without a cease-fire. However, he instructed his negotiators to announce a "unilateral truce" (perhaps to save face). This step was rejected by the rebels as a maneuver to make them cease fighting "prematurely."

By contrast, when the insurgents are losing an internal war, they—rather than the central government—may desire a cease-fire. In the last year of the Greek civil war, the Communist insurgents twice proposed a cease-fire. In May, 1949, after they had suffered severe setbacks, the Communists proposed a cease-fire and general elections under United Nations supervision. Even though the Communists omitted their earlier demands for the withdrawal of the American and British military aid missions, the Greek government rejected the proposal, insisting that the Communist army had to give up its arms.

In October, 1949, after the Communist insurgency in Greece had been suppressed, the Greek Communist leaders made a second "cease-fire" proposal from their exile in Bulgaria. It appears that this move was forced upon them by Stalin for his own purposes, perhaps to free the remnants of the Greek Communist army for actions against Tito.[3]

In the first year of the Korean war, the Communist side, despite its severe military setbacks, maintained demands that would have given it a complete victory. For instance, on December 13, 1950, the Soviet ambassador to the United Nations, Jacob Malik, voted against an Asian-Arab cease-fire resolution in the United Nations General Assembly because it did not stipulate "the withdrawal of foreign troops, which is the first condition for a settlement of the Korean question." On December 22, 1950, Communist China's Foreign Minister, Chou En-lai, rejected a cease-fire, arguing that when the United Nations troops had crossed the 38th parallel they "completely destroyed and consequently wiped out forever the demarcation line of political geography. . . . We firmly insist that as a basis for negotiating a peaceful settlement . . . all foreign troops must be withdrawn from Korea . . . [and that] the U.S. forces must be withdrawn from Taiwan." By June, 1951, all these demands had been dropped and the Communist Chinese representatives sat down together with North Koreans to negotiate a cease-fire on the very demarcation line that allegedly had been "completely destroyed and wiped out forever"—nay, on a line slightly less favorable to the Communist side.

According to some indications, the Chinese Communists, in changing their position, seemed to have responded in part to Stalin's pressures. In part, the dismissal by President Truman of General Douglas MacArthur (who had advocated attacks against China) may have reassured the Chinese by indicating that American objectives were more limited than they had anticipated. That is to say, an internal change in the American government may have strengthened the forces on the Communist side who advocated that the Korean war be ended by accepting a stalemate.

Once negotiations on a cease-fire had started, however, they dragged on for over three years, while the fighting continued. The United Nations side was anxious to reach a settlement because it continued to suffer high casualties, and the United States, in particular, was concerned about the fate of the many American prisoners in Communist hands.

Early in the Korean truce talks, agreement was reached on a specific new boundary separating North Korea from South Korea, the so-called demarcation line. Through this partial settlement, the Communists clearly limited the risks that further fighting might entail for them. It was as though they placed a floor under their future losses. The agreed demarcation line made it unlikely that a United Nations advance would result in anything but a temporary loss of territory for the Communist side, since the United Nations side would have found it difficult to renege on its agreement. At the same time, the agreement to repartition Korea in this manner (slightly to the north, on the whole, of the prewar boundary) meant that a settlement had in essence been reached. The residual disagreement chiefly concerned the Communist demand that the United Nations side return prisoners from Communist China by force, even against the prisoners' own wishes.

Washington and the United Nations team at the negotiations were aware that early agreement on the demarcation line could protect the Communists from further loss of territory. But Washington nonetheless instructed the American commander of the

United Nations forces, General Ridgway, to accept this line in the interest of reaching an early agreement, and because the line met the basic American position on the maintenance of its defenses. However, Washington was badly disappointed in its expectation that the rest of the agenda would be settled within a month.[4]

In 1951 the Communists mounted several offensives against the demarcation line, and these had to be fought back with high losses. Tentative American plans provided that if the negotiations broke down or became "hopeless," an offensive might be launched north of the line. Plans for an advance to the Wonsan-Pyongyang line and even as far as the Yalu were brought up to date, but General Ridgway thought that under the prevailing circumstances neither of these offensives would be worth the casualties they would cause. In the following year as well (1952), the United Nations side abstained from a major offensive northward. In June, 1953, the Communists finally agreed to voluntary repatriation of the prisoners of war, essentially on the original United Nations terms; and a month later the armistice was finally concluded.

The outcome of these protracted negotiations, thus, was that the Communist side yielded on the prisoner of war issue. The terms of the truce settlement, as a whole, differed little from what the United Nations side had proposed quite early in the talks. But the ability of the Communists to procrastinate cost the United Nations side dearly. More Americans were killed during the two years of truce negotiations than during the first year of the war before negotiations started. Among all the United Nations forces, fatalities during the negotiating period were about double those suffered previously.

Had Washington anticipated these high casualties, it might have tried harder either to obtain an immediate cease-fire when the negotiations began, or else to maintain a threat of territorial losses against the Communist side by denying it the protection of an agreed demarcation line. These suggestions, however, are easily offered with the advantage of hindsight. The President and his chief advisers were under pressures at home to end the war, and he could not count on retaining his domestic support for a harder bargaining strategy even though its purpose would have been to save American lives.

It is still unclear what finally brought the Communists around. Internal changes on the Communist side may have played a key role, and perhaps the change of administration in Washington had some effect as well. The Soviet leadership was in turmoil after Stalin's death. A new foreign policy, less hostile toward the West, was in the making in Moscow. In the United States, Eisenhower's election enabled the American government to examine policies for coping with the stalemated war that were bolder than those of the Truman administration during its last year in office. Word went out from Washington through various channels that the United States might decide to expand the war geographically and perhaps even introduce nuclear weapons. It is possible— though we cannot know for sure—that the Communist leadership sensed a grave risk and decided it was not worth taking. All the Communist side had to do to ward off the danger of escalation was to accept a certain loss in prestige by consenting to a voluntary repatriation of the prisoners, which would mean that most of the captured Communist troops would opt to stay on the United Nations side.[5]

Conflicts among the allies on the United Nations side, particularly between the Americans and the South Koreans—the principal parties—also contributed to delays in the negotiations. After June 8, 1953, when the details of the prisoner of war issue were agreed upon, another seven weeks passed before the armistice was signed on July 27. During this period, the American authorities had to bring South Korea's President Syngman Rhee around to the settlement.

In the end, Syngman Rhee must have realized that the truce settlement gave him about as much as the United States was willing to fight for. The dependence of his forces on American support provided the pressure, the security treaty with the United States the inducement, to abandon his more "hawkish" position. The American pressures were all the more effective because of the fact that Rhee could not find a political faction in Washington, or among the American commanders in the field, that would take his side in the internal American debate.

General Clark, the American commander of the UN forces, gave Rhee no encouragement in his recalcitrance. He was sympathetic to Rhee's concern for American support in the future, hence he favored the mutual security treaty—a suggestion later approved by Washington. Apart from this concession, however, Clark and his staff took an almost harder position against Rhee than did President Eisenhower and his advisers in Washington.[6]

By contrast, in many other wars, a "hawkish" faction within a country has been able to make common cause with a "hawkish" ally against a compromise peace. Or, putting it in less derogatory language, those who are opposed to defeatism can receive political support in their struggle at home from an ally whose interests would have to be violated if the fighting were to be brought to an end.

In World War I, for instance, a separate peace between the Western Powers and Imperial Austria was prevented by the more "hawkish" allies on both sides. On the side of the Central Powers, as we have seen, leading Austrian officials sacrificed the interests of their own country (and virtually betrayed the peace plans of their Emperor) in preference to betraying the alliance with Germany. On the side of the Western Allies, the "hawks" in the French government (and to a lesser extent those in the British government) managed to kill the Austrian peace feelers in 1917 by permitting their Italian ally to cast a veto. For it was the Italian annexationists who would have had to be disappointed if Austria were drawn away from Germany. (As in 1940, the Italian armies were unable to collect the spoils by their own effort.) Totally embracing the Italian position, the French ambassador in

Rome worked particularly hard to dissuade his government from making peace with Austria.[7] As so often happens, the ambassador *to* a country becomes the ambassador *from* that country.

During the negotiations on Japan's surrender in World War II, a struggle between "hawks" and "doves" in Washington and among some of the Allies dangerously reinforced the "hawkish" faction in Tokyo. For one thing, the "unconditional surrender" formula of the Allied Powers gravely handicapped those within the Japanese government who wanted to end the war, but who could not prevail without guarantees that the Emperor would be allowed to remain.

In discussing the difficulties of utilizing cryptographic information about the enemy, we have seen how peace feelers from the Japanese Foreign Office sent to the Japanese ambassador in Moscow were intercepted and decoded by the American intelligence service. Among those top American officials who were privy to such information, some saw it as evidence in support of the more conciliatory policy they favored, a policy which, instead of insisting on the unconditional surrender formula, would have informed the Japanese that they could keep their Emperor.

The intricate bureaucracy in Washington, however, was directing the final phases of the gigantic military effort in the Pacific and at the same time preparing for the President's summit conference in Potsdam with Churchill and Stalin. The policy-making mechanism of wartime Washington was not a flexible instrument for negotiating war termination with another government, especially one whose own peacemakers were constantly on the verge of being overthrown by a clique of fanatic officers. The "hawks" in Washington, unwittingly teaming up with the "hawks" in Tokyo, almost prevented Japan's surrender. The Chinese and Australian governments lent support to the "hawks" in Washington by their reluctance to offer any guarantees for the preservation of the Emperor. In addition, some State Department officials were opposed to what they considered appeasement and feared an adverse reaction among the American public, particularly if the effort failed. Officials on the military side of the American government, by contrast, were in the vanguard in easing Japan's surrender terms.[8]

The negotiations for Japan's surrender in 1945 illustrate how the technicalities of communicating with the enemy government can become a serious obstacle to peacemaking. The problem is not that the central leadership in a government cannot communicate with the enemy's central leadership—there are always many channels open for that (except, perhaps, in a nuclear war). The difficulty arises when the central leadership is divided, and when the "doves" (who may be in control of the foreign office or one of the military services) must first establish that the enemy's terms can be lived with and then try to win support for their peacemaking effort. This has to be done unobtrusively, even furtively; and it can easily come close to a conspiracy where the line between treason and patriotism appears blurred.

For the Japanese surrender in 1945, the feat of communicating with the United States was just barely accomplished. The Japanese "doves" were twice set back in their effort to establish contact with the United States. In March, 1945, the Japanese naval attaché in Switzerland got in touch, through Allen Dulles, with the American wartime intelligence service, but the attaché's reports regarding possibilities for an early settlement were waylaid by the "hawks" in Tokyo. A second, more official Japanese attempt sought to enlist the Soviet government as mediator. This approach was somewhat fumbled, in particular as it was combined with an attempt to interest the Soviet Union in a rapprochement with Japan. The Soviets would not exert themselves as peacemakers and passed the messages on to the United States in a cursory fashion, without any constructive suggestions for Tokyo.[9] In choosing a mediator, one must be sure to choose a party that wants to see the war shortened. Why should Stalin have wanted to see the war

SEARCH FOR AN EXIT

against Japan terminated before he could enter it himself to collect his spoils?

The Korean war, in contrast to World War II, was a typically stalemated war. Neither side was militarily defeated, and neither side could have developed a realistic plan to overwhelm the enemy. The fighting might have lasted even longer had it not been for the fact that a repartitioning of Korea could be so easily arranged, and that this partition practically restored the *status quo*. In conflicts that are predominantly civil wars, however, outcomes intermediate between victory and defeat are difficult to construct. If partition is not a feasible outcome because the belligerents are not geographically separable, one side has to get all, or nearly so, since there cannot be two governments ruling over one country, and since the passions aroused and the political cleavages opened render a sharing of power unworkable.

The Communist insurgency in Greece ended in 1949 with the national government in control of the entire country and the Communist party banned. In an earlier phase in the Greek civil war, the Communists could have chosen to give up their armed struggle while surviving as a party more or less free to use nonviolent means of opposition. But in the internal conflict among Communist "hawks" and "doves," Zachariades, the "hawkish" proponent of an all-out fight against the government army, prevailed. Thereafter, Tito's break with Stalin deepened the dissension among the Greek Communist leadership. In the end, Zachariades managed to blame his opponents within the Communist party for the defeat (which he had engineered) and denounced them as "traitors"—the Communist equivalent of the stab-in-back legend.[10]

WHEN THE FACTS ARE TOO PAINFUL

Closely related to the internal debate about tactics and objectives of peace negotiations is the debate about military prospects. And this debate in turn interacts with the disagreements about war aims. To bring the fighting to an end, one nation or the other almost always has to revise its war aims. This revision is stimulated by a reevaluation of the military prospects, but not in so direct a linkage nor in so logical a fashion as a rational approach to national policy would dictate.

The debate on military prospects concerns well-guarded professional domains from which the public, and sometimes even major parts of government officialdom, are excluded. While this cloak of secrecy and arcane specialization rarely is abused to influence government policy through deliberate lies—though that happens sometimes—it often tends to cover up highly fallacious reasoning. Quite apart from distortions and logical fallacies, estimating the military prospects in a war, as we have seen, is subject to large uncertainties. Given these uncertainties, and given the lack of agreed criteria and procedures for estimating the outcome of further fighting, both the "hawks" and the "doves" can always find military estimates to support their case.

Both sides in the internal debate often pick some isolated statistics out of the welter of information, instead of trying to justify their position on the basis of an over-all evaluation. Selecting from the same set of data, those who contend that the struggle must continue may cite high casualties of the enemy forces; those in favor of a compromise peace may stress large enemy reinforcements. The former group may use indicators of declining enemy morale in support of their argument; the latter group may counter by pointing out that the enemy has increased his attacks. And so forth.

It requires sternly disciplined detachment for the military and civilian leaders to separate their political views about the advisability of further fighting from their evaluation of the military prospects. "Every expert is a human being," wrote A. J. P. Taylor, "and technical opinions reflect the political views of those who give them. Generals and admirals are confident of winning a war when they want to fight; they always find decisive arguments against a war which they regard as politically undesirable." (Taylor recalls that at the time of Mussolini's attack on Abyssinia, when British naval leaders were opposed to intervention, the British cabinet received a report from its naval advisers to the effect that the British navy in the Mediterranean, even if reinforced by the entire Home Fleet, was no match for the combined Italian navy and air force. Yet, in World War II the British navy beat the Italians despite worse odds than those of 1935.) [11]

Clearly, in the midst of a war those who sort out and interpret the flood of raw information, reported by a great variety of sources, have plenty of room to misinform everybody. As Lord Asquith, British Prime Minister in the early years of World War I, once put it, the military in the War Office "kept three sets of figures, one to mislead the public, another to mislead the cabinet, and a third to mislead itself." Even among evaluations by historians with all their leisurely hindsight and with no problems of prediction, wide discrepancies can be found. To cite only one instance (reported by the historian Leon Wolff), the British War Office in 1922 published figures on the British casualties for the 1917 battles in Flanders half as large as those reported in 1948 in the official British history. The difference might have been due to the opposite biases of a favorable and a critical attitude toward General Haig's strategy. [12]

Deliberate misrepresentation is not the only sort of distortion that affects domestic debates on war termination. Government leaders and bureaucrats introduce other distortions in opposing each other, not so much to confound their domestic opponents, but rather to permit their continued functioning within a single government. In deciding how a war should be ended, those who rule the country must choose among basic national goals. The more unfavorable the outcome of the war, the more deeply must these choices cut into fundamental values and threaten to create paralyzing divisions in domestic politics.

To make peace may require that the nation get rid of its

leader. But the leader, in seeking advice from his ministers on how to end the war, cannot ask for a frank debate on his own political demise. Or, to make peace may require the abandonment of war aims for which men are still being asked to die. If the leaders who wish to argue for such a peace denigrate these war aims, they would be asserting that the men at the front are dying in vain. To make peace may require disbanding the existing army (or conversely, letting it rule the country). But the civilian and military leaders in deciding how to end the war cannot have a frank debate on how to abolish each other. If these factions argued their positions until they reached the logical conclusion, their differences would become so personal that they could not work together as members of the same government. Delicately balanced coalition governments are particularly constrained in coping with these fundamental decisions.

Hence the peculiar fashion in which government leaders— according to so many historical records—reason with each other when they must decide how to end a war that goes badly. Instances abound where energetic and relatively intelligent men, accustomed to positions of leadership, deal in a most inconclusive, almost unworldly fashion with questions of life and death for their country.

Among the many devices by which domestic factions avoid joining the essential, but all too touchy issues, two stand out. One is to debate the timing of a crucial decision without ever discussing whether or not the move should be made at all. (This is, as well, a standard bureaucratic procedure.) The other device consists in concealing the concrete differences behind a rhetorical contest: instead of deciding which policy, given all its domestic implications, would be least bad for the country—for that is often the choice—the leaders pretend that the choice lies between "treason" and "honor." Perhaps they even persuade themselves that the choice can be cast in such simplistically moral terms.

In World War I, Germany's internal debate about starting the unlimited submarine campaign became largely a debate about timing, just as Japanese debate in the decision to attack Pearl Harbor shifted to the question of "when" before really settling the question "whether." As we have seen, the advocates of the German campaign never precisely explained how such an escalation would help to terminate the war; they merely asserted that it would "break England's back" in five months. Their opponents, however, failed to attack the central weakness of the proposal; they only dared to argue about its timing. The German Chancellor, Bethmann-Hollweg, did not feel he had the power to say to the military that their proposal was no good at all; his political strength permitted him only to say "not yet." Typically, when Field Marshal Hindenburg pressed for the final decision (January 8, 1917) he stressed the urgency rather than the merits of the move: "It must be," he exclaimed; "we are counting on the possibility of war with the United States, and we have made all the preparations. It cannot get worse. The war must be shortened by the use of every means." [13]

When it appeared that the tide had turned against them in September, 1918, the German military leaders feared that the enemy would break through the German lines. On October 1, Ludendorff suddenly demanded of the civilian leaders that they immediately approach President Wilson with a peace proposal. On that day a new German government was just being formed. The civilian leaders did not wish to approach the Allied Powers in such haste, even though they favored peace negotiations more strongly than the military. Grudgingly, Hindenburg (Ludendorff's senior colleague in Germany's military leadership) agreed the peace move might be made the following day, if the new government were ready by then: "If, on the other hand, the formation of the Government is in any way doubtful, I consider the dispatch of the declaration to the foreign governments as imperative for tonight." [14]

Thus, the civilian-military debate focused entirely on the

question of timing, and the more fundamental questions regarding the likely enemy conditions and the necessary German concessions were hardly discussed. These issues were too painful, and the various factions in Germany were too deeply divided on them. Indeed, while Hindenburg would scarcely wait overnight for the peace proposal to be made, he still wanted the annexation of Longwy-Brie, a rich industrial area that had been part of France for 250 years.[15]

Within the next two weeks, the German military reversed themselves again, feeling more confident about their ability to stem the Allied advance and at the same time disillusioned by President Wilson's cool reaction to the first German peace offer. A long conference was held on October 17 between the civilian and military leaders. Now the debate almost got around to the unmentionable issue of what Germany would have to offer to obtain peace:

General Ludendorff: I am under the impression that before accepting the conditions of this note, which are too severe, we should say to the enemy: Win such terms by fighting for them.
The Imperial Chancellor: And when they have won them, will they not impose worse conditions?
General Ludendorff: There can be no worse conditions.
The Imperial Chancellor: Oh, yes, they can invade Germany and lay waste the country.
General Ludendorff: Things have not gone that far yet.

Thus, the military leadership remained unable, or unwilling, to face the painful facts.

After some further prodding by the civilians, Ludendorff made it clear that he wanted an armistice "to permit an orderly evacuation" by his troops and to obtain "a respite of at least two or three months." [16] What he feared was that the enemy wanted to make the resumption of hostilities impossible for Germany. In this fear he was, of course, right. The position of the Allies was far too strong for them to entertain proposals that would grant the Germans a respite.

This conference touched on the basic cleavage between the civilian and the military leaders. But one cannot debate regicide in the presence of the king. The military wished to protect the army and maintain their hold on German politics; should the Allied conditions preclude this, "there could be no worse conditions." The civilians, on the other hand, could envisage greater disaster (some, indeed, would have considered a trimming of the military's powers not altogether a disaster); there could be greater casualties and more destructive damage to the German economy, and, in the end, the central German government might be destroyed.

A few days later, Hindenburg telephoned the Chancellor: "But even if we should be beaten, we should not really be worse off than if we were to accept everything at present. The question must be asked: will the German people fight for their honor, not only in words but with deeds, to the last man and thereby assure themselves of the possibility of a new existence." [17] This is the kind of rhetorical question that "must be asked" only if one wishes to avoid discussing the real policy choices.

In 1943–44, the Finnish government took over a year and a half to draw the obvious and painful conclusion from the clearly perceived prospect of eventual military defeat. There was no argument about the military situation. As early as February 9, 1943, the head of military intelligence reported the unfavorable military outlook to the members of parliament, warning them of the need to make peace with the Soviet Union. Opportunities were not lacking for acting upon this knowledge. In July, 1943, the Soviet legation in Stockholm inquired whether Finland would be willing to enter into peace negotiations. Also around that time various declarations in favor of negotiations were voiced by members of the Finnish Social Democratic Party. Again in November, the Soviets in Stockholm informed the Finns that they were willing to receive an authorized person to discuss peace proposals; and Sweden, attempting to be helpful, offered to provide substitutes for the food supplies that Germany would, of course, no longer furnish when Finland left the war. At last, in March, 1944, a Finnish delegation went to Moscow to learn the Soviet peace terms. As we have seen, only in August after Mannerheim became President was the Finnish government able to take the final step.[18]

One can only wonder where the Finnish leaders thought their country was headed, given the course to which they adhered throughout these months. What calculations—if any—could they have been making while they persisted in following the road to disaster? The Finnish leaders, however, were probably governed less by mistaken calculations than by reluctance to confront painful facts. They had to recognize that their decision to start the war had been a dreadful mistake; they had to order a yet unbeaten army to give up national territory.

The task of terminating an unsuccessful or stalemated war not only breaks alliances apart and splits governments. It can also turn a decisive government leader into a Hamlet. When the Pétain government was about to be installed in Bordeaux, several of the French leaders who had been firmly opposed to negotiations with Hitler became hesitant. Prime Minister Reynaud seemed to collaborate in his own ouster, no longer knowing which road he should take. The day before the fall of the Reynaud government, Admiral Darlan spoke of his plans to leave with the French fleet and continue the fight alongside England; yet the following day he accepted the post of minister of the navy under Pétain.[19]

The task of bringing an unsuccessful war to an end demands such a soul-searching reordering of objectives that many government leaders respond to it with failure of nerve. Strong men may lapse into strange indecisiveness, precisely at a time when their nation most urgently requires firm decisions. Conversely, leaders who have remained in the background may pick up the reins and guide their country out of the war. Emperor Hirohito, during a few critical weeks in the summer of 1945, managed to assert himself as the supreme decision-maker and thus saved the integrity of Japan. He has been fully vindicated in envisaging "recovery and a brighter future" for Japan when he announced to his wavering war cabinet his decision to accept the surrender terms that the Allied Powers had laid down at the Potsdam Conference.

The case of the Italian dictator, Mussolini, is a particularly striking example of how having to face the painful facts of a lost war can paralyze government leaders. Mussolini—once so decisive or even impulsive in all his actions—became a man who was fighting with himself. During the last six months of his reign in Rome, he was unable to choose between pursuing the war, on the one hand, and making the best out of Italy's surrender, on the other. Instead, he kept wavering between incompatible courses of action. He continued to oppose the betrayal of Germany, while he knowingly supported those who worked for this betrayal. He held on to power, while he willingly took the road that led to his ouster; indeed, he almost prepared his ouster himself.

In January, 1943, he chose Vittorio Ambrosio, an anti-German, as his new chief of staff—a remarkable appointment. Mussolini must have known that Ambrosio was anti-German. As it turned out, Ambrosio was instrumental in arranging Italy's surrender to the Allies half a year later; his role was more active and more effective than Badoglio's.[20] Then in March, as we have seen, Mussolini began to pay more attention to military estimates while at the same time obfuscating the facts. He began to realize that the Axis had to make political sacrifices and was willing to jettison some war aims—not yet his own aims, but Hitler's. In a long letter to Hitler he wrote: "In my opinion Russia cannot be annihilated, not even through the improbable intervention of Japan, given the enormous distances. The Russian chapter must therefore be liquidated in one way or another." But Mussolini did not have the courage of his convictions and failed to insist on a change in policy even on this—from his point of view eminently sound— idea of trying to make a separate peace with Russia. In his next meeting with Hitler he failed to press this proposal seriously.[21]

On July 14, 1943, four days after the Allied landings in Sicily, Ambrosio wrote a memorandum to Mussolini analyzing the military situation. Ambrosio dared to hint broadly at the inevitable political conclusion:

If one cannot prevent the setting up of [a second front by the Allies], it will be up to the highest political authorities to consider whether it would not be expedient to spare the country further sorrow and ruin, and to anticipate the end of the struggle, seeing that the final result will undoubtedly be worse in one or more years.

And a day later, Ambrosio is reported to have told Mussolini:

. . . the war is lost because Germany is not in a position to bring us immediately the necessary aid. . . . Italy cannot accept the use of her territory with no hope of salvation, as the outer defense of the Reich. . . . Germany is, on the other hand, for geographical and industrial reasons able to hold out against the enemy for another year, perhaps longer. . . . One must not hesitate to cross the ditch before it is too late. And if the Germans want to make Italy their battlefield, one should not altogether exclude the possibility that Italy should fight against these allies who have systematically failed in their word.[22]

Thus, Mussolini kept as his chief of staff a man who advised him to break with his ally and to capitulate to the enemy, and who even hinted that Italy should be prepared to wage war against its ally. Yet he was unable to take any step in that direction. When Mussolini met Hitler for the last time before his fall from power, Ambrosio privately kept pressing him to get out of the war. Mussolini is reported to have responded to the urgings of his chief of staff as follows:

Perhaps you think that this problem has not been consciously in my mind for a long time. . . . I admit the hypothesis: to detach ourselves from Germany. It sounds so simple: one day, at a given hour, one sends a radio message to the enemy. . . . But with what consequences? The enemy rightly will insist on a capitulation. Are we ready to wipe out with one stroke a regime of twenty years and the results of a long bitter effort, to admit our first military and political defeat, to disappear from the world scene? . . . What attitude will Hitler take? Perhaps you think that he would give us liberty of action.[23]

In everything that he said in those days, Mussolini was not arguing that he would remain on "the world scene," or that the military situation was better than his chief of staff told him. But under the stress of mounting catastrophe he could not muster the determination to choose a policy that would have corresponded to this military situation. So painful had the facts become that he could no longer face them.

ENDING WARS BEFORE THEY START

"IF INDEED war should break out," Khrushchev wrote President Kennedy during the Cuban missile crisis of 1962, "then it would not be in our power to stop it, for such is the logic of war. I have participated in two wars and know that war ends when it has rolled through cities and villages, everywhere sowing death and destruction." [1] With forceful language, Khrushchev warned that if the United States and Russia should become embroiled in a war, their governments would no longer be able to control the conflict. Then, in the same famous letter, he offered to pull out his missiles and thus provided President Kennedy with the critical offer for settling the crisis.

Those with power to start a war frequently come to discover that they lack the power to stop it. In those intensely frightening days of the Cuban missile crisis, Khrushchev recalled not only the vast destruction his country suffered in World War II but also— from the time when he was a young man—Russia's tremendous losses in World War I. He must have remembered how, in both these wars, Russia was locked in, once the fighting had started. In 1917 the newly formed Bolshevik government had to surrender a third of Tsarist Russia to German control in order to extricate itself from threatened annihilation. And in World War II, after

1943 Stalin never had the choice of making another separate deal
with Hitler, to let the "Capitalists" fight each other, as he did in
1939. Stalin did not try to make such a deal; but had he wanted
to, that option would have been closed to him. Despite Germany's
mounting defeats both in the East and in the West, Hitler rejected
Mussolini's advice, in the winter of 1942 and the spring of 1943,
to make a separate peace with Russia.

Indeed, Hitler was locked in, too. He argued—with good
reason—that such a peace would not permit him to shift his
forces against the West, since the Russians would merely take advantage of a respite and then attack his Eastern front.[2] How could
he have relied on a new agreement with Stalin, having himself
broken the Nazi-Soviet nonaggression pact? (Hitler broke that pact
by launching his fierce offensive in 1941, even though Stalin had
made every effort at that time to avoid war with Germany.)

Whatever the obstacles to an arrangement that would have
prevented war, the use of violence itself engenders new obstacles
to the reestablishment of peace. Fighting sharpens feelings of hostility. It creates fears that an opponent might again resort to violence, and thus adds to the skepticism about a compromise peace.
As we have seen, more is expected of a settlement because both
the government and the people will feel that the outcome of the
war ought to justify the sacrifices incurred. In addition, various in-
stitutional forces will compound the difficulties of making peace.
Wars are not like peacetime military maneuvers that can be halted
when the results are in.

When antagonists have nuclear armaments, the need to prevent war is, of course, particularly compelling. At the same time,
prevention might be easier because nuclear weapons simplify—in
a terrible sense—the task of estimating the results of warfare. The
destruction they cause can outweigh every other aspect of fighting,
and such destruction is more certainly calculable than that resulting from conventional weapons. Hence, statesmen are forced, as
never in the past, to consider how a war would end before becom-

ing entrapped in those processes through which fighting, once started even at a low level of violence, tends to prolong itself. The stark knowledge of the danger of nuclear warfare intensifies the search for ways to prevent it.

In the present as in the past, there are basically two ways to prevent war: by eliminating the sources of conflict that would lead a nation to resort to the use of arms, and by rendering the use of arms so unattractive that a nation would rather tolerate existing conflicts or frustrations than start a war. The former approach relies on conciliation and concessions and might be accompanied by disarmament; the latter relies on dissuasion and deterrence and might be supported by certain arms control arrangements. Each of these approaches, to succeed, requires that the antagonists hold certain views of how a possible war might end. Those with power to start a war must expect that the ending would be worse than what their nation would have to concede, or tolerate, to preserve the peace. This condition is well understood; indeed, it is what modern deterrence strategy is all about.

Less clearly recognized by most modern strategists and officials who shape military policies is another condition necessary for the prevention of war. Expectations regarding the outcome must not only look worse than the price for peace; they must also clearly govern all the decisions and dynamics through which military violence might be unleashed. It is not enough that those who can deliberately start a war should at no time come to believe that their nation, or their "cause," would be better served by going to war than if peace were maintained. For even if this condition is met, it will not be sufficient if wars can be started by technical accidents, or started by leaders who fail to think coherently how the fighting will end, or who, in some perverse stubbornness, no longer care if it ends in disaster for their own country.

Many wars in this century have been started with only the most nebulous expectations regarding the outcome, on the strength of plans that paid little, if any, attention to the ending. Many began inadvertently, without any plans at all.

FROM APPEASEMENT TO DETERRENCE

> *If we handle Hitler right, my belief is that he will
> become gradually more pacific. But if we treat him as a
> pariah or mad dog we shall turn him finally and irre-
> vocably into one. I would feel confident if it were not
> for the British press or . . . that section of it which is
> inspired either by an intelligentsia which hates Hitler
> . . . or by alarmists by profession and Jews.*
> —SIR NEVILE HENDERSON *British Ambassador in
> Berlin, February, 1939*

Among the many different forms of conflict between nations, some
create a far greater risk of war than others. Attempts by one country to change the existing situation are particularly critical if a
rival nation sees in them a threat to its prestige, its political interests, or—above all—its military security. Just what kind of political or military arrangements are viewed as "vital" by a nation is
an elusive question, for much of the reasoning about "vital interests" tends to be circular: those interests are "vital" that are considered as such.

Evaluations of changes in the status quo that would worsen a
nation's military security can be somewhat less abstract than evaluations concerning threats to "national prestige" or "political" interests. Yet, even for this purely military dimension, the existing
level of hostility toward the nation that would gain in military
power is all-decisive. For instance, in 1935 the incorporation of
the Saar into Germany appeared threatening to France. In 1955–
56 this incorporation was viewed in Paris as part of a policy of
reconciliation. Or today, a build-up in the Libyan air force seems
threatening to Israel, a build-up in the Iranian air force does not.

Since foe might become friend and allies might become enemies, the long-term effects of political, military, or territorial
changes must remain uncertain. It is therefore often a difficult task
for a government to weigh whether to oppose a change that appears threatening—if necessary by going to war—or whether to

attempt to live with it, or even turn it into an occasion for reconciliation. If Austria, Russia, and Germany, the principal powers
that first went to war in 1914, could have foreseen how that war
would end for them, each government would have preferred to accept almost any change in the status quo that its opponent might
have insisted on. Whether or not concessions by any of these powers might in fact have prevented World War I is one of the hypo-
thetical questions about history that can never be answered. What
is certain, though, is that concessions and conciliation were
scarcely tried.

The fact that World War II was not prevented cannot, by
contrast, be blamed on a lack of concessions to the powers seeking
to change the status quo. On the contrary, the prewar policy of
England and France—the status quo nations of that time—has
been criticized for precisely the opposite error, too much appeasement. It is the bitter experience of this period that turned the
word "appeasement" into a derogatory term. Prior to the late
1930s, "appeasement" did not mean feeding the appetite and
power of an aggressor, but pacifying through concessions a conflict
that threatened to erupt into a war. For the present era, it is critically important to understand how appeasement can succeed or
fail, without being swayed by false lessons from the 1930s.

It has become customary to refer to political or territorial
concessions that one disapproves of as a "second Munich." This
criticism implies that the demands or pressures of an expansionist
adversary should be resisted, as at Munich in 1938 the British and
French supposedly should have resisted Hitler's annexation of
major parts of Czechoslovakia. What concerns us here, of course,
is not so much the historical details of the 1930s, but the more
general question of how policymakers, in choosing whether or not
to "appease," weigh the alternative risks, and especially the possible outcomes, of a war.

When Prime Minister Chamberlain met with French Premier
Daladier, shortly before the Munich Conference, to work out a
common Anglo-French negotiating position, Daladier favored a

tougher stand than Chamberlain. Chamberlain gave some disquiet-
ing estimates of Western military weakness to support his position
that the Western Powers should minimize the risk of war. He
asked what assurances France had received from Russia, and
pointed out that the British government, for its part, had received
very disturbing news about the probable Russian attitude. "It
would be a poor consolation," he added, "if, in fulfillment of her
obligations, France attempted to come to the assistance of her
friend but found herself unable to keep up her resistance and
collapsed." [3] After the collapse of France in 1940, Chamberlain
could have recalled this remark as a warning that sadly came true.

During these summit talks, Daladier, on the other hand, appears to have judged that his policy would look best if one ignored
the question of how the fighting would continue, should war break
out, let alone how it would end. According to the British memorandum of this remarkable conversation,

Chamberlain asked what should be done if Herr Hitler refused [the
compromise proposed by Daladier]. Daladier said that in that case
each of us would have to do his duty. Chamberlain thought we should
have to go a little further than that. . . . Daladier said he had no further proposal to make. Chamberlain thought we could not fence about
this question. We had to get down to the stern reality of the situation.
. . . Hitler, instead of temporizing any further, would interpret [Dala-
dier's proposed reply as] a rejection of his proposals and say he would
march into Czechoslavakia. Chamberlain wished to know what the
French attitude would be in such an event. Daladier replied that Hitler
would have then brought about the situation in which aggression would
have been provoked by him. Chamberlain asked what then. Daladier
thought each of us would do what was incumbent upon him. Chamberlain asked whether we were to understand from that that France
would declare war on Germany.[4]

At this point Daladier finally pointed to the French mobilization
measures, and after further probing by Chamberlain as to what
France would really do, he argued that the German system of for-
tifications was less solid than Hitler had indicated and that, therefore, once French troops had been concentrated, an offensive

should be attempted by land against Germany, and that "as re-
gards to the air, it would be possible to attack certain important
military and industrial centers."

Daladier may have thought that appeasement was unnecessary because deterrence would work. Hence, he felt more comfortable with his vacuous military plans. Three weeks before Munich,
he told the British ambassador he was "convinced that, if Hitler
could only be made to realize that German aggression on Czechoslovakia means a general war, he would abstain." Questioned by
the British ambassador whether in the event of war there would
not ensue a stalemate in the Maginot and Siegfried lines, Daladier
denied this and said he was convinced he would be able to undertake "a series of limited offensives." [5]

Those who try to avoid appeasement by threatening war must
think about how such a war might end. It is now generally understood that statesmen and military planners should not rely on deterrence without a careful evaluation of the precise kind of
retaliation or response that their side could mount and how a potential aggressor might view such a threat. One can sympathize
with British lack of confidence that Hitler would be deterred by
the prospect of the French army carrying out "a series of limited
offensives."

There is much to be criticized in the Munich deal, particularly the sanctimoniousness of the Chamberlain government about
its attempt to appease Hitler—at the expense of Czechoslovakia.
But one argument in favor of Munich is that it gave England and
France another year in which to prepare for war (although Germany too gained a year). Another argument in favor of Munich is
that it showed up Hitler's next aggressive move as such a flagrant
violation of a treaty he had just signed, and which had granted
him vast concessions, that it strengthened popular support in the
West for fighting Nazi Germany in what turned out to be a long
and costly war.

It has also been claimed that the time to oppose Hitler by

force was in 1936, when France and Britain decided in favor of the first major appeasement by letting Hitler remilitarize the Rhineland Zone. The rationale behind this claim is that Germany was then still weaker than France—as contrasted with 1938 or 1939. But for varied reasons no military action was taken by the French when Hitler sent his troops into the Rhineland, in violation of the Versailles treaty. A contributing factor seems to have been the hesitant attitude of the French military. Asked to evaluate military possibilities, the head of the French army, General Gamelin, argued that a deep movement into the Rhineland would require general mobilization. This the political leaders were unwilling to impose on the nation, fearing—and probably rightly— lack of support among the French people.[6]

The French military may have exaggerated the need for mobilization prior to taking any action. But critics of the appeasement in the Rhineland crisis may have overestimated the ease with which Hitler and his regime could have been toppled. Let us assume the French government had decided to intervene by force. Given the lack of popular support for a war to destroy Nazism by marching on to Berlin, and given that none of the responsible officials had even envisaged such a war, this intervention would have had to be in the form of small attacks across the border.[7] The French High Command had long made plans for such limited local action, namely, the reoccupation of the Saar and the seizing of "pawns" in the Rhineland. The military planners viewed these measures as "acts of coercion." [8]

What if this "coercion" had failed to coerce? Having witnessed how Hitler was able to hold on to power in circumstances far more hopeless than a French reoccupation of the Saar and perhaps a few border towns, one cannot simply assume that these "acts of coercion" would have put an end to Nazism.[9] How would that war then have ended?

In France and England, neither the governments nor the public were psychologically prepared in 1936 to wage a major war against Germany in order to eliminate the Nazi regime. It was precisely the failure of the attempted appeasements during the Rhineland crisis and at Munich that convinced increasingly larger segments of governmental and public opinion in the Western Powers that Hitler had to be eliminated, even at the price of a costly and dangerous war. Indeed, as late as 1939, by no means all the senior officials were so convinced.

In trying to draw lessons from the appeasement of the 1930s one should not overlook the possibility that the outcome could have been even worse. If Hitler could have rallied German patriotism, his army would even by that time have been strong enough to checkmate the limited invasions that the French High Command had planned. After a while, the French government, unwilling to escalate, might have had to negotiate a humiliating withdrawal which could have strengthened French defeatism, and left as a residue the lesson never again to intervene beyond France's borders.

Twenty years later, when Egypt nationalized the Suez Canal, Prime Minister Eden feared his country would be acceding to a "second Munich" (or a "second Rhineland") unless it took some drastic action to reverse the Egyptian moves. The fear of becoming trapped by a policy of appeasement led Eden, as we have seen, into his initiation of a war without any plans for ending it.

Dangers of appeasement are not avoided merely by "inoculating" the aggressor, as it were, through the puncture of a small-scale military intervention. Readiness is required—preferably backed up by plans and preparations—to carry a war to a successful conclusion, should the initial intervention fail to induce the opponent to abstain from (or to rescind) his aggression.

The most dangerous phase in following a policy of "appeasement" arises when initial concessions fail to pacify the opponent. The difficulties are several. The switch to a policy of deterrence may be attempted without the requisite military preparations, precisely because in a period of appeasement such preparations usually lack popular support and would seem inconsistent with a stance of conciliation. Hence, the opponent cannot be confronted by a military capability threatening him with losses severe enough to outweigh his expected gains from aggression.

A further difficulty—especially for a democratic government —is to achieve both an official consensus and effective public agreement on the point at which appeasement (in the nonderogatory sense of the word) must be abandoned and any further demands by the opponent rejected, backed up by the threat of war. There is likely to be room for internal disagreements about whether "appeasement" should not have been abandoned some time ago, or whether conciliation and concessions should not be tried a little longer. In the new confrontation with the "aggressor," those who opposed appeasement before will only see proof that their government's policy had been mistaken. Those officials who argued in favor of concessions in the past, on the other hand, may urge that their policy should be continued, in the hope that it will yet be crowned with success.

Thus, a few hours after Hitler took over the rest of Czechoslovakia in March, 1939, in violation of the Munich agreement, the British ambassador in Berlin, Sir Nevile Henderson, wrote to the British Foreign Secretary, Lord Halifax: "Hitler has gone straight off the deep end again. It has all come very unexpectedly . . . and the extremists have again won the day and all one can hope is that they will eventually regret it. What distresses me more than anything else is the handle which it will give to the critics of Munich." [10] That is to say, the good ambassador was distressed not so much that appeasement might objectively have failed as that his critics could claim it to have failed.

This difficult transition from a policy of concessions (or appeasement) to a policy of deterrence often produces a lack of clarity regarding which concession is meant to be the last one. This uncertainty will make it more likely that the last appeasement will fail to pacify the opponent and that the subsequent attempt at deterrence will fail to dissuade him from seeking further gains.

Such a lack of clarity regarding how far concessions were meant to go can be seen in British policy on the issue of the Suez Canal. In the early 1950s, Britain still kept troops stationed along the Suez Canal in accordance with a treaty with Egypt dating back to 1936. The Egyptians exerted mounting pressure against the British forces on their soil, including sporadic guerrilla warfare. But in 1954 the British government agreed to evacuate its troops and through this concession—which some British parliamentarians thought shameful "appeasement"—vastly improved its relations with Egypt. Premier Nasser called it "a good agreement on which we can start at once to build a new basis of relationship with Britain and the West." A possibly prolonged war between British troops and Egyptian guerrillas had thus been prevented.

However, the improvement in Anglo-Egyptian relations came to naught. The British still had a major interest in the Suez Canal Company, whose lease from Egypt for the exclusive right to operate the canal had another fourteen years to run. The British did not at all envisage, as a further concession to Egypt, permitting the nationalization of the Canal Company. Yet, not a word was said in the agreement of 1954 that would have drawn a line between the first concession—the troop withdrawal—and what to Egyptian nationalists must have seemed the logical second concession—the nationalization of the Canal Company. According to the text of the 1954 agreement, the Egyptian government merely promised to uphold the 1888 convention guaranteeing the freedom of navigation of the canal, a promise which it kept as far as British navigation was concerned. It did not promise to refrain from nationalizing the Canal Company. Thus, when that step followed in 1956, Prime Minister Eden thought appeasement had failed again as in the 1930s, and since it was too late for deterrence he decided to go to war.

Anthony Eden seems to have drawn the wrong lesson from the appeasement of the 1930s (which he had opposed with much foresight and courage). What ensured the failure of efforts to prevent the outbreak of World War II was not only the fact that a man like Hitler could not be appeased but also the fact that the Western Powers sought to deter Hitler with the threat of war, yet had no military plans for bringing such a war to an end. After the violation of the Munich agreement in March, 1939, Britain and France abandoned all attempts to appease Hitler and switched to a coordinated policy of deterrence. Thus, the purpose of their hurried guarantees to Poland and Rumania was not to establish an effective military alliance for fighting Germany but to put an end to appeasement by deterring Hitler from committing further aggression.

Yet, having outflanked these alliances by his pact with Stalin, Hitler attacked Poland undeterred by the threat of war with the Western Powers. He recognized that they lacked the capability to defeat Germany. He apparently hoped their threat would turn out to be a bluff—he had not even asked his military staff to plan any offensive operations against the West.[11] It is not that the Western Powers failed to warn Hitler. For instance, in May, 1939, the British Foreign Minister instructed his ambassador in Berlin to warn General Keitel, the German commander in chief, that the British and French would honor their guarantee; and even though "the German government may regard such an eventuality with equanimity . . . [the Western Powers] would be quite certain to triumph in the end. In other words, if Herr Hitler provoked a war over Danzig, it would result not only in the destruction of the Nazi regime, but also very probably in the final collapse of the great-German Reich." [12]

This warning of how the war would end came, of course, completely true. Yet, not even the British could have foreseen in 1939 how by honoring their guarantee to Poland they would start a war that would bring the United States and the Soviet Union into the conflict on their side and hence end so catastrophically for Germany. Since the British prediction could at best have been based on intuitive foresight, how could it have deterred Hitler who always prided himself on his own intuition?

THE FUTURE OF DETERRENCE

> *I must make one admission, and any admission is formidable. The deterrent does not cover the case of lunatics or dictators in the mood of Hitler when he found himself in his final dug-out. That is a blank.*
> —WINSTON CHURCHILL

> *We will never capitulate, never. . . . We might be destroyed, perhaps; but we will drag a world with us— a world in flames.*—ADOLF HITLER

Deterrence must reinforce appeasement to prevent war. While appeasement alone, or "conciliation" alone (to use a word without derogatory connotation), has often succeeded in initiating an enduring peace, it has sometimes failed tragically. A main source of failure has been that the nation which sought to pacify its opponent could not—or neglected to—establish a line and convince the opponent that there his gains must stop.

A military relationship among nations based on deterrence alone may also prevent war for a long time. But if deep hostilities and the roots of sharp conflict persist, continued reliance on deterrence cannot foreclose all avenues that might lead to war. For the success of deterrence depends on whether the many individuals who hold keys to war or peace think coherently—or think at all —about how fighting, if started, would come to an end.

Since 1945, the growing knowledge about nuclear weapons and their steadily mounting destructive potential have revolutionized the thinking and planning on deterrence among all major powers. During the first decade following the destruction of Hiroshima and Nagasaki, there was still some hope in Western countries that nuclear war could be prevented by abolishing nuclear weapons through international controls. But during the second nuclear decade, statesmen and the public at large became increasingly resigned to the notion that to prevent nuclear war the powers possessing these weapons had to rely on deterrence. Thereafter, the leading views on military policy—particularly in the United States—began to interpret what seemed to be the only strategy that *could* prevent nuclear war as the strategy that reliably *would* prevent it. Nuclear deterrence became accepted as a relationship between the two major nuclear powers that should be preserved by "stabilizing" it.

The essential ideas of mutual deterrence were given their classic statement to the world at large in 1955 by Winston Churchill in his famous speech on the "balance of terror." [13] The ideas had been developed over several years, of course, by military strategists in the United States, England, and perhaps other countries, too. Since then, a few important refinements have been added, but the basic structure of thought has remained unchanged. Thus, barely ten years after the first atomic explosion, the grand strategy for the nuclear age became a commonly accepted view—at least in the Western world.

Churchill foresaw in 1955 that within a number of years both the United States and the Soviet Union would be "capable of inflicting crippling or quasi-mortal injury on the other." Expounding the comforting features of mutual deterrence, Churchill added: "It does not follow, however, that the risk of war will then be greater. Indeed, it is arguable that it will be less, for both sides will realize that global war would result in mutual annihilation. . . . After a certain point has been passed, it may be said, the worse things get, the better."

Indeed, things did get worse. Advances in nuclear technology permitted ever more destructive weapons; stockpiles of nuclear warheads kept constantly growing; long-range bombers, which offered at least hours of warning and reflection time, were superseded by missiles that, once fired, could not be recalled, and whose warheads were first numbered by tens, then by hundreds, and now by thousands; and meanwhile, the number of nuclear powers has grown from three to eight or more. Yet, the worse things got, the better—or so many people came to think. As the physical reality of this destructive potential became more and more lethal for the whole world, the state of mind of responsible people became rather less anxious.

In the 1950s and early 1960s, there was still some talk about what might happen after nuclear deterrence had failed. British strategists, for instance, raised the question whether, after a nuclear attack, some kind of "broken-back warfare" was a possibility; and Soviet statesmen debated whether Communism might survive a nuclear war or whether Capitalism and Communism would both perish. Today, nuclear war is seen in only two acts: the initial attack that is to be deterred, and the retaliatory strike that constitutes the deterrent. The possibility of later phases in a full-scale nuclear war, let alone how such a war would end, is no longer a question that people wish to debate.

Given this state of mind, all the troubling problems somehow have to fit under the Great Deterrent. For instance, some fifteen years ago so-called tactical nuclear weapons in Europe were viewed by NATO strategists as providing a way to repel a massive conventional attack. But today, the thousands of these enormously destructive weapons are intended—if any coherent thought is devoted to them at all—exclusively as a means to extend the Great Deterrent in a way that would prevent that massive conventional attack.

China's build-up of her nuclear arsenal, uninhibited by the Non-proliferation Treaty, is another development that could raise doubts about the "balance of terror." Yet, according to prevailing Western opinion, China will be guided, as her nuclear missiles increase in numbers, by the same deterrence strategy that seems to govern the American-Soviet relationship and into which Great Britain and France have easily fitted. In debating the merits and shortcomings of an antiballistic missile system, United States Senators and Congressmen found little difficulty in extending America's reliance on the Great Deterrent from the customary bipolar relationship to a triangle of relationships so as to include China.

At first blush, this exclusive reliance on the Great Deterrent to cope with the dangers of nuclear war has impressive arguments in its favor. First, it seems to have worked in the past. Since 1945, not a single nuclear weapon has been used to cause destruction. Second, nothing better that seems realizable has been suggested. International controls and disarmament that would abolish the threat of nuclear weapons never have appeared within sufficiently reliable grasp to be chosen as an alternative policy by the governments of the great powers. And defenses against nuclear as a substitute for exclusive reliance on deterrence have been rejected, either as too unpromising on technical grounds or as jeopardizing the "stability" of deterrence.

Deterrence aims to convey with stark clarity to those with power to start a nuclear war that whatever their expectations as to how a war might end, the final settlement would be overshadowed by the losses incurred from the retaliation. Accordingly, nearly all effort goes into ensuring that the adversary must expect this all-decisive loss. In a sense, important progress in strategic thinking since World War II is to be found in the fact that a nuclear attack and the opponent's retaliation would be inherent in the very design of plans for a first strike. Accordingly, the term "first strike capability" has come to mean a capability that can prevent retaliation or reduce it to "acceptable" levels.

However, the exclusive reliance on nuclear deterrence appears less comforting—indeed, deeply troubling—if its many limi-

122 ENDING WARS BEFORE THEY START

tations are taken into consideration. Deterrence must be viewed as a strategy that seeks to bring the initiation of a war in which nuclear weapons are to be used into the closest possible relationship with the war's outcome. But this attempt is endangered by many of the perverse processes that have so often prevented governments from either fully understanding or effectively planning the termination of wars. Several problems with frightening implications jeopardize the reliability of nuclear deterrence strategy.

A basic difficulty can be recognized even if one thinks within the narrow confines of the theoretical scheme that is used to justify universal reliance on the Great Deterrent. It remains questionable whether the execution of a so-called retaliatory strike can serve national interests, once it has failed as a threat. To be sure, the "retaliatory" strike might be carried out anyhow, because there might no longer exist the degree of control, the time, or the political will to stop in mid-course the elaborately planned procedures for striking back. But it is also possible that the national leadership would recognize that the execution of a "retaliatory" strike, once the threat of it had failed to prevent war, was a new question to be decided on its merits.

In such a cataclysmic situation, the question might well be answered on emotional grounds—that "retaliation" is to be carried out for the purpose of "revenge." Conversely, however, the national leadership might give deeper thought to the fate of its devastated country and realize that what remains of the nation's future would largely depend on how the war, which had just begun, would come to an end. If anything was to be salvaged, the withholding of the undestroyed nuclear weapons might seem the wiser choice. President Nixon has raised this problem quite explicitly:

Should a President, in the event of a nuclear attack, be left with the single option of ordering the mass destruction of enemy civilians, in the face of the certainty that it would be followed by the mass slaughter of Americans? Should the concept of assured destruction be narrowly defined and should it be the only measure of our ability to deter the variety of threats we may face? [14]

While this complexity has long been recognized by strategists,[15] its implications are more serious than has generally been admitted. For deterrence to be effective against a *deliberately* planned attack—and only against this can it be effective—the would-be aggressor must give more thought than the defender as to how the war would end. Or more precisely, the would-be aggressor must be prudent enough to calculate how the victim's retaliation would deny him a desirable outcome in all circumstances. Yet, to be deterred he must believe that the victim possesses precisely the opposite trait; that is, he must expect the victim to retaliate in a reflex fashion reckless of cost or consequences. History teaches us that these characteristics tend rather to be reversed between aggressor and defender, and that many an aggressor has counted on this being so.

Another danger affects the reliability of deterrence in preventing nuclear war: the possibility that a conventional war between *nuclear* powers might not be ended before one side resorts to the use of nuclear weapons. The more prolonged and fierce the fighting, the less feasible a return to the prewar relationship with the enemy. Even if the governments on both sides were determined initially to avoid any use of nuclear weapons whatsoever, difficulties in bringing the fighting to an end might weaken this determination. As the suffering among the population increased and as government leaders on both sides, for internal political reasons, felt a growing need to obtain an outcome better than a mere stalemate, the fighting might rise to increasingly higher levels of violence.

To make matters worse, each government would be most reluctant to reach a settlement by accepting a partial defeat, lest it thereby seem to signal to its enemy that it feared nuclear war more than the enemy, and thus invite further aggression. This desire to avoid a military defeat, hence, would soon overshadow the limited objectives for which the fighting might have started. Once this point had been reached, deterrence could be maintained to prevent the use of nuclear weapons only if the government leaders on both sides—while trying to bring the conventional fighting to a halt—continued to give first priority to national survival over all other war aims. In many wars in the past, government leaders have lost sight of this priority.

One may be tempted to brush aside the danger of escalation to a nuclear war in the belief that nuclear powers, sensing this risk, would always compromise their local conflicts, rather than fight each other directly even with conventional weapons. Unfortunately, since the open fighting along the Sino-Soviet border in 1969, this argument has lost some of its force.

Other types of dangers must be considered in asking whether "the balance of terror" can prevent nuclear war. In the United States, a great deal of emphasis, not only in strategic analysis and doctrine but in actual military programs, has gone into making this balance more secure against a possible Soviet temptation deliberately to launch a massive strike for the purpose of disarming the American retaliatory capability. Such a temptation could develop, it is feared (particularly in time of crisis), if the American retaliatory forces seemed vulnerable to a Soviet surprise attack.[16] Important as such efforts are, they can cope with only some of the dangers of nuclear war. There are other potential avenues leading to the outbreak of nuclear war that cannot be blocked by strengthening deterrence.

That is to say, nuclear war might break out through processes against which deterrence is of no help. Perhaps foremost among these is the risk that nuclear missiles or bombers might be sent against their target by accident: through some unanticipated technical malfunctioning, through unauthorized actions by personnel in control of a delivery system, or through a combination of the two. Not only could such an accidental launch itself result in unprecedented destruction, but it might lead to a full-scale nuclear war. (The launch might trigger a massive return strike mistakenly intended as "retaliation" for a deliberate attack, or the nation where the accidental launch originated might decide on a preemptive follow-up attack to blunt such "retaliation.")

Obviously, a threat of retaliation cannot directly prevent such accidents or unauthorized acts. Indirectly, mutual deterrence affects the risk of accidents in two contrary ways. On the one hand, to be sure, the enormity of the disaster that could in turn befall a country where the accident originates will act, it is to be hoped, as a constant spur to all nuclear powers to maintain effective safeguards. But on the other hand, reliance on deterrence can have a perverse effect on the accident risk.

The very function of "retaliatory" forces requires that they can be launched after the fabric of the nation has been torn apart. Nuclear powers, to ensure that they cannot be disarmed through a surprise attack, may keep elements of their forces ready to be released instantly and may decentralize the arrangements for controlling these weapons. Such measures tend to increase the probability that technical accidents or unauthorized acts will lead to the launching of nuclear weapons. At the same time, exclusive reliance on the Great Deterrent could make ever more likely the dreadful possibility that such an accident would lead to a full-scale nuclear war.

Historians have blamed the military staffs of the European powers before 1914 for rigging their mobilization schedules and planned responses to an adversary's mobilization in such a way that limited military intervention by one power in an accidentally triggered local conflict automatically engulfed all those nations, within a few weeks, in one of history's most destructive wars. But the situation today may be even worse, or could soon become so. Today, some government leaders and strategists have become so possessed by the idea that everything ought to depend upon deterrence that they press for military preparations that make the preparations of 1914 look prudently flexible and rationally cautious.

For example, a senior American Senator recommended—in earnest, apparently, not as a macabre jest—that American intercontinental missiles should be launched "immediately" upon warning of a Soviet attack, "without any fiddling around about it, even without asking the computer what to do," and even if the warning indicated a "light attack, one that could be detected." What might appear as a "light attack" within the available minutes and "without asking the computer what to do" could, of course, be an accidental launch or even a false signal. The critical period during which this uncertainty could not be resolved would be precisely the moment when the United States government should bend every effort to prevent an accidental use of nuclear weapons (let alone a false radar warning) from automatically triggering a full-scale nuclear war.[17] This was not the view of one individual only. Some other American Senators and proponents of certain strategic policies have also recommended that the United States adopt a policy of launching its missiles upon the receipt of radar warning, or at least immediately after the first nuclear explosion confirmed such warning.[18]

Apart from such public advocacy, military officers in various countries may demand in private that their governments adopt the same or similar doctrines. The Soviet doctrine of "preemption" —which somehow seems to envisage striking a swift nuclear blow at the enemy *before* he can complete (or even start?) his attack— gives rise to fears that the American advocates of launch-on-warning may find their counterparts among some of the more "hawkish" Soviet military officers.[19] And if one side gives the appearance of adopting a launch-on-warning policy, the other side might also feel compelled to institute faster launch procedures—creating something of an "arms race" in reducing the safeguards against accidental war.

The hazard is elusive. It is inherent not only in the weaponry but also in the command and alert procedures, during periods of tension as well as during the months and years of quiet waiting. Only those unsafe conditions can ever be corrected that have been discovered or thought of.[20] The weapons systems that make up "the balance of terror" are so exceedingly complex, both in their internal structure and in how they relate to the entire military posture, that it seems questionable whether any group of military or technical experts could ferret out all ramifications to discover every danger spot. Indeed, the notion of a weapons "system" tends to be misleading, in that it implies a closed set of interrelated parts all designed to serve a common purpose. An aeronautical engineer who might have studied the safety of aviation "systems" some years ago would not have included the hijacker within his "system." In the past, aircraft design took no account of this risk.

The possibilities for the outbreak of a nuclear war that cannot be blocked by mutual deterrence are broader than the risk of an accidental weapons launch. Deterrence is based on the premise that people in control of nuclear weapons wish their country to survive. Yet there has never been a period in history without men acquiring positions of power who were willing to die, and to see others die, for causes that they themselves invented and which were espoused by only a few of their henchmen. In several countries the political process is such that leaders can come to the top who consider it a virtue, or perhaps part of their "revolutionary" creed, to live dangerously. *Vivere pericolosamente* was one of Benito Mussolini's favorite slogans.

One is reminded of the almost total lack of forethought that Hitler and his military staff gave to the decision to declare war against the United States. During the first hours after this decision had become accomplished fact, Hitler's chief of staff, General Jodl, telephoned from Berlin to one of his officers in General Headquarters: "You have heard that the Führer just declared war against the United States? It is now the task of the Staff to examine in which direction, Far East or Europe, the United States are likely to turn the bulk of their forces. Only after that will further decisions have to be taken." The officer responded: "Certainly, this examination seems much needed. Since the war against the United States was hitherto not supposed to be considered and since we therefore have made no preparations for this examination, the task has to be undertaken at once." General Jodl responded: "See what you can do. When we meet tomorrow, we can discuss this further."[21]

By relying on nuclear deterrence, the major powers presumed a certain harmony between their own strategic views and those of other nuclear powers. While Americans and Russians during the Cold War sought to understand each other's views, they have treated their differences as errors in thinking to be corrected in long disarmament conferences and numerous informal talks by patient and persistent education. In the meantime, the views of both sides have undergone considerable evolution; and now, indeed, each side sees the errors in ideas that it firmly defended in the past.

There was a time, for instance, when American strategists recommended civil defense measures to protect urban populations, when American arms control proposals argued that reductions in conventional forces should precede limitations on nuclear weapons, and when the official United States position belittled the risk of nuclear accidents and accidental war. Since then, all these positions have been reversed at least once, and in some cases twice. During the same period, the Soviet government has turned upside down its position of demanding that reductions in nuclear armaments (if not their abolition) had to precede arms control measures designed to make mutual deterrence more "stable"; and Soviet spokesmen, who only a few years ago argued that active defenses against missiles should be increased rather than limited, now take the opposite view.[22]

One might take comfort in the notion that these revisions are salutary, reflecting only the fact that each nuclear power is bringing its strategy up to date. But updated toward what goal? In view of the harsh divergence in political outlook among the Western Powers, Russia, and China, can one expect that their strategic views will all converge in a single school of thought? Besides, every strategic theory contains large uncertainties that can be settled by intuition only, not by facts. In coping with these uncertainties, therefore, strategists and statesmen may have to listen to the echo of their own ideas.

It has been pointed out that in the physical sciences new theories gradually replace old ones, again to be rejected as false. Knowing that the view of the world keeps changing in the exact sciences, we should be all the more ready to anticipate, and given our present precarious situation even welcome, intellectual revolutions in the "soft" science of strategy and arms control. The theory of stable deterrence"—with all it limitations—is in itself unlikely to remain stable.

It is also commonplace in human affairs that men continue to labor on major undertakings a long time after the ideas upon which these efforts were based have become obsolete. In World War I, the premises on the strength of which the principal belligerents entered into the fighting were proven wrong within a few weeks; yet the mutual slaughter proceeded for four more years. Since one cannot constantly reexamine one's premises, it is easy to overlook a growing discrepancy between reality as it changes, and the old intellectual foundations of an ongoing policy. Such discrepancies have become particularly disastrous in the conduct of wars, where the fighting continued even though the objectives and ambitions which gave rise to it had long been overtaken by events. However, policies that seek to prevent wars can likewise be based on ideas that have been outpaced by technology or rendered obsolete by the internal momentum of national military programs.

The language in which the strategy of deterrence has been discussed obscured the fact that this strategy is based on a scheme of totally unprecedented cruelty. Various abstractions and metaphors—which remain necessarily (and fortunately) untested by reality—help to insulate the design against the wrath of the innocents who are its target. Owing to these metaphors, a scheme that would have been rejected as abhorrent in the

Dark Ages by kings and common people alike, appeared to reflect the humane ideals of modern civilization:

"Mutual" deterrence ("mutual" sounds like an arrangement that distributes benefits evenly)

will be "stable" (that is, of unfailing continuity)

if both "sides" (and later three "sides"?)

maintain nuclear capabilities so that a "potential aggressor" (as if all wars started because somebody so decided)

must expect that his attack would lead to a nuclear "exchange" ("exchange" as with commodities in foreign trade?)

resulting in the "assured" destruction ("assured" seems to convey comfortable certitude)

of "his" cities (are the cities to be destroyed necessarily the property of those who would start a nuclear war?).

After the end of the Cold War, strategic experts tried to bring nuclear strategy up to date. The collapse of the Warsaw Pact had dissolved the massive military confrontation in the center of Europe, which was the principal reason for America's nuclear buildup and had stimulated the Soviet buildup as well. The profound political changes in Russia convinced nuclear planners, on both sides, that they no longer faced an unbridgeable enmity. Gone were the reciprocal fears and hatreds between the "Bolsheviks in the Kremlin" and the "Wall Street capitalists." Yet four decades of Cold War calcified bureaucratic habits and accumulated an immense collection of armaments. This double legacy could not be easily overcome. The military bureaucracies had to find a strategic rationale for the Cold War weaponry, and the easiest solution was to continue the old rationale.

It took another political earthquake to wake up the nuclear planners in Washington and Moscow. After 9/11, almost everyone realized that Russia and the United States no longer were archenemies. In December 2001, President Bush informed President Putin of the U.S. withdrawal from the Cold War treaty banning missile defense. This step, at last, ended the strategy of "mutual assured destruction." Americans had been divided on this issue. Some had wanted to scuttle this treaty years ago;

others wanted to keep it and predicted the U.S. withdrawal would provoke a new nuclear arms race. These predictions were mistaken. Putin
took the U.S. withdrawal in stride. Like President Bush, he understood
the Cold War was over and that both Russia and America had to confront
new threats.

One of these new threats is terrorism. It is exceedingly difficult to
end a war in which the enemy leader can enthrall masses of followers
who are eager to carry out terrorist attacks as suicide bombers. For
decades, the best intentioned efforts to end protracted terrorist wars have
failed again and again—in Kashmir, Chechnya, Israel-Palestine, Sri
Lanka, Colombia. And for our war against al Qaeda, a negotiated peace
is not even being mentioned by either side. The more deep-seated the enmity, the more difficult it is to find an exit from the fighting. Indeed, the
nation that is the victim of terrorist tactics might have no option for ending the war short of accepting the enemy's demands or annihilating him.
For instance, if the terrorist warfare in Afghanistan continued beyond
2004, despite the successful elections, total annihilation of the Taliban
might be the only option that will bring peace. If the day ever comes
when a global terrorist organization uses nuclear weapons, the attacked
nation might face the abhorrent choice between surrender and inflicting
genocide on the populations that bring forth the terrorists. We must find
a way to end such wars before they start.

NOTES

PREFACE TO THE REVISED EDITION

1. Interview with Foreign Minister Eduard Shevardnadze, reported in *Gazet Wyborcza*, Warsaw, October 27, 1989.

2. Richard Nixon, *No More Vietnams* (New York: Arbor House, 1985), p. 45.

3. Reflecting on this episode, Weinberger wrote later: "My own feeling was that we should not commit American troops to any situation unless the objectives were so important to American interests that we had to fight, and that if those conditions were met, and all diplomatic efforts failed, then we had to commit, as a last resort, not just token forces to provide an American presence, but enough forces to win and win overwhelmingly." Caspar W. Weinberger, *Fighting for Peace* (New York: Warner Books, 1990), pp. 159–60.

4. Accuracy and discrimination in the use of offensive armaments is important not only for these political reasons but also to make military forces more efficient. The fewer munitions that miss the target, the fewer the attacks needed and the less demanding the required logistics support. In sum, relying on indiscriminate destruction is not a prudent strategy: it is a wasteful use of military capabilities; it is normally a grave mistake in seeking to end a war; and as a threat, indiscriminate destruction also makes an unreliable deterrent. Cf. *Discriminate Deterrence*, Report of the Commission on Integrated Long-Term Strategy, Fred C. Iklé and Albert Wohlstetter, Co-Chairmen (U.S. Government Printing Office, 1988).

5. Supplying the Afghan resistance with American-made anti-aircraft missiles (the Stingers) may have played a critical role in convincing the Soviet military that the costs of their Afghan war were becoming too high. See Fred Barnes, "Victory in Afghanistan: The Inside Story," *Reader's Digest*, December 1988, pp. 87–93.

134 134

6. I remember well, while serving as Undersecretary of Defense during these years, that the danger of such a Soviet escalation was carefully considered by those of us in Washington who dealt with American aid for the Afghan resistance fighters.

7. I offered some preliminary thoughts on this point in "Nuclear Strategy: Can There Be a Happy Ending?," *Foreign Affairs*, Spring 1985, pp. 824–25.

CHAPTER I. THE PURPOSE OF FIGHTING

Abraham Lincoln's statement was quoted approvingly by Lloyd George on December 19, 1916, when he delivered a vehement speech in the House of Commons against suggestions for starting peace negotiations. General Douglas MacArthur expressed his view in an address to a joint meeting of Congress, April 19, 1951, after his dismissal as commander of the United Nations forces in the Korean war.

1. Nobutaka Ike, *Japan's Decision for War: Records of the 1941 Policy Conferences* (Stanford, Calif.: Stanford University Press, 1967), p. 133.

2. *Ibid.*, pp. 139–40 and 153.

3. *Ibid.*, pp. 202–4.

4. André Beaufre, *The Suez Expedition 1956* (New York: Praeger, 1969), pp. 29, 49, 54–55, 80, 136; Kenneth Love, *Suez: The Twice-Fought War* (New York: McGraw-Hill, 1969), pp. 392, 424, 458.

5. Beaufre, *Suez Expedition*, p. 143.

6. Fritz Fischer, *Griff nach der Weltmacht: Die Kriegszielpolitik des Kaiserlichen Deutschland, 1914–1918* (Düsseldorf: Droste, 1961), p. 218.

7. One reason why World War I lasted so long and ended so disastrously for Germany was that her military leaders insisted on extensive territorial annexations. The rationale for these annexations was largely the idea that they would deter the enemy powers from challenging Germany to another war, or should deterrence fail, provide Germany with a stronger position for fighting such a war. See Klaus W. Epstein, "Development of German-Austrian War Aims in the Spring of 1917," *Journal of Central European Affairs* (April, 1957), pp. 24–47.

8. Bismarck, *Die Gesammelten Werke*, XIII, 209–16 (speech to the Reichstag, January 11, 1887); and Gerhard Ritter, *Staatskunst und Kriegshandwerk* (Munich: R. Oldenbourg, 1959), I, 325–26.

9. Charles Seymour, *The Intimate Papers of Colonel House* (Boston: Houghton Mifflin, 1928), III, 45-46.

10. Studies on the making of foreign policy, by and large, have tended to neglect the internal processes and "bureaucratic" factors within governments, such as rivalries among government agencies and military services, career motivations and personal ambitions of subordinate government offi-

cials. A pioneering study that stressed these "bureaucratic" factors is Richard E. Neustadt, *Presidential Power—the Politics of Leadership* (New York: John Wiley, 1960). Recent studies which address this aspect more explicitly are Graham T. Allison, "Conceptual Models and the Cuban Missile Crisis," *American Political Science Review* (September, 1969), pp. 689–718; and Morton H. Halperin, *Bureaucratic Politics and Foreign Policy* (Washington, D.C.: Brookings Institution, 1971).

CHAPTER II. THE FOG OF MILITARY ESTIMATES

The quotation from Karl von Clausewitz is from the first part of his book *On War*, which he wrote between 1816 and 1830. (*Vom Kriege* [Berlin: Dümmler, 1857], I, 76.)

1. Owing to some duplication and independence in wartime intelligence research, a minority view could assert itself within the American intelligence community that evaluated Japan's economic strength more correctly. *U.S. Strategic Bombing Survey: The Effects of Bombing on Japan's War Economy* (Washington, D.C.: U.S. Government Printing Office, 1946), pp. 69–72.

2. Burton H. Klein, *Germany's Economic Preparations for War* (Cambridge, Mass.: Harvard University Press, 1959), p. 201 and *passim*.

3. Walter G. Hermes, *Truce Tent and Fighting Front* ("United States Army in the Korean War" [U.S. Army, Office of Chief of Military History, Washington, D.C.: U.S. Government Printing Office, 1966]), pp. 331–32, 296.

4. Anthony F. Upton, *Finland in Crisis, 1940–1941: A Study in Small Power Politics* (Ithaca, N.Y.: Cornell University Press, 1964), p. 218 and *passim*.

5. Anatole C. Mazour, *Finland Between East and West* (Princeton, N.J.: Van Nostrand, 1956), pp. 157, 158.

6. The Earl of Birkenhead, *The Professor and the Prime Minister* (Boston: Houghton Mifflin, 1962), p. 262. Cf. Albert Wohlstetter, "Strategy and Natural Scientists," in Robert Gilpin and Christopher Wright, eds., *Scientists and National Policy Making* (New York: Columbia University Press, 1964); and C. P. Snow, *Science and Government* (Cambridge, Mass.: Harvard University Press, 1961).

7. Fred Charles Iklé, *The Social Impact of Bomb Destruction* (Norman. Oklahoma: University of Oklahoma Press, 1958), pp. 175–76.

8. Gordon Smith, "R.A.F. War Plans and British Foreign Policy: 1935–1940" (Ph.D. thesis, M.I.T., Cambridge, Mass., 1966). On the circumstances that led from controlled to indiscriminate air warfare between Great Britain and Nazi Germany, see F. M. Sallagar, *The Road to Total War: Escalation*

in World War II (Santa Monica, Calif.: The Rand Corporation, 1969), especially chap. VII.

9. Carnegie Endowment for International Peace, *Preliminary History of the Armistice* [official documents published by the German Reichskanzlei] (New York: Oxford University Press, 1924). Original German text in Herbert Michaelis, ed., *Ursachen und Folgen: Vom deutschen Zusammenbruch 1918 und 1945 bis zur staatlichen Neuordnung Deutschlands in der Gegenwart* (Berlin: Wendler, 1959), II, 281.

10. One can also identify the signals (at least in retrospect) that might have served to distinguish these Chinese warnings from the blustering language the Chinese use routinely. Allen Whiting, *China Crosses the Yalu* (New York: Macmillan, 1960), pp. 109–12.

11. For one of the most useful and exhaustive accounts of the history of cryptography and code breaking see David Kahn, *The Code Breakers: The Story of Secret Writing* (New York: Macmillan, 1967). On Pearl Harbor, see Roberta Wohlstetter, *Pearl Harbor: Warning and Decision* (Stanford, Calif.: Stanford University Press, 1962). On the British use of the German telegram to Mexico in World War I, see Barbara Tuchman, *The Zimmerman Telegram* (New York: Macmillan, 1966), and Kahn, *The Code Breakers*, chap. 9. Reference to Germany's breaking of British codes in World War I is made in Deutsche Nationalversammlung, 1919–20, Untersuchungsausschuss über die Weltkriegsverantwortlichkeit, *Stenographische Berichte über die Öffentlichen Verhandlungen des 15 Untersuchungsausschusses. . . nebst Beilagen* (Berlin, 1920), I, 358–59. See also Kahn, *The Code Breakers*, chap. 10.

12. On the German interception of Churchill-Roosevelt conversations, see Kahn, *The Code Breakers*, pp. 556–57; and Albert N. Garland and Howard McGraw Smyth, *Sicily and the Surrender of Italy* ("U.S. Army in World War II, The Mediterranean Theatre of Operations" [Washington, D.C.: U.S. Government Printing Office, 1965]), p. 287. On the many tactical military advantages derived from code breaking in World War II, see Kahn, *The Code Breakers*, chaps. 14–17.

On German successes in reading foreign diplomatic communications prior to and during World War II, see David Irving, ed., *Breach of Security: The German Secret Intelligence File on Events Leading to the Second World War* (London: Kimber, 1968). See especially pp. 35–40 for a discussion of the difficulties governments have in using information derived from cryptographic intelligence.

13. For these Japanese telegrams see U.S. Department of State, *Foreign Relations of the United States: The Conference of Berlin (Potsdam), 1945*, I, 873–83, and II, 1248–64. For Stalin's communication, see *ibid.*, II, 460. For Stimson's memorandum, *ibid.*, II, 1266. James Forrestal also knew of these messages, noting in his diary on July 13, 1943, that they offered "the

first real evidence of a Japanese desire to get out of the war." (Walter Millis, ed., *The Forrestal Diaries* [New York: Viking Press, 1951], p. 74).

14. Hugh Gibson, ed., *The Ciano Diaries, 1939–1943* (Garden City, N.Y.: Doubleday, 1946), p. 553.

15. F. W. Deakin, *The Brutal Friendship: Mussolini, Hitler and the Fall of Italian Fascism* (New York: Harper & Row, 1962), pp. 212–13.

16. Those who are unaware of the uncertainty of seemingly simple and clear-cut economic statistics should consult Oscar Morgenstern, *On the Accuracy of Economic Observations* (Princeton, N.J.: Princeton University Press, 1963). The balance sheet of a firm, foreign trade statistics, census figures, etc., are all subject to various types of important errors.

CHAPTER III. PEACE THROUGH ESCALATION?

The exchange between Lyndon B. Johnson and Douglas MacArthur took place in the hearings on General MacArthur's dismissal: U.S. Senate, Committee on Foreign Relations, *Military Situation in the Far East* (Washington, D.C., 1951), p. 211.

1. A study that features the metaphor of an escalation ladder prominently is Herman Kahn, *On Escalation: Metaphors and Scenarios* (New York: Praeger, 1965). But Kahn, too, recognizes that the "rungs" are often artificial dividing lines across a continuum. He also reminds us that participants in a war may have different perceptions of the various levels in the use of military power. The motives for expanding or limiting wars and the interaction with the enemy in limiting violence, conventional as well as nuclear, are dealt with in Morton H. Halperin, *Limited War in the Nuclear Age* (New York: John Wiley, 1963).

2. Fritz Fischer, *Griff nach der Weltmacht: Die Kriegszielpolitik des Kaiserlichen Deutschland, 1914–1918* (Düsseldorf: Droste, 1961), p. 394.

3. Karl E. Birnbaum, *Peace Moves and U-Boat Warfare: A Study of Imperial Germany's Policy Towards the United States, April 18, 1916–January 9, 1917* (Stockholm: Almquist & Wiksell, 1958), pp. 317–18.

4. The memorandum of the German navy recommending the submarine campaign is reproduced in full in hearings held by the postwar German government regarding the reponsibilities for Germany's defeat in World War I: Deutsche Nationalversammlung, 1919–20, Untersuchungsausschuss über die Weltkriegsverantwortlichkeit, *Stenographische Berichte über die öffentlichen Verhandlungen des 15. Untersuchungsausschusses . . . nebst Beilagen* (Berlin, 1920), Vol. II, Part V, pp. 225–82.

5. The tonnage sunk in February, 1916, was 540,000; in March, 574,000; in April, 881,000; in May, 596,000; and in June, 700,000. Henry Newbolt,

Naval Operations: History of the Great War Based on Official Documents, by Direction of the Historical Section of the Committee of Imperial Defense (London: Longmans, Green & Co., 1931), IV, 359, 370, 385; V, 42, 58.

6. William H. Beveridge, *British Food Control* (London: Oxford University Press, 1928), p. 346. And N. B. Dearle, *An Economic Chronicle of the Great War, for Great Britain and Ireland* (Carnegie Endowment for International Peace, "British Series in Economic and Social History of the World War" [London: Oxford University Press, 1929]).

7. Birnbaum, *Peace Moves*, pp. 178–79. Germany, Nationalversammlung, *Stenographische Berichte*, Vol. II, Part I, *Aktenstücke zur Friedensaktion Wilsons, 1916–17*, p. 179.

8. Germany, Nationalversammlung, *Stenographische Berichte*, Vol. II, Part V, p. 229.

9. *Ibid.*, I, 345. Johann-Heinrich von Bernstorff, *Deutschland und Amerika: Erinnerungen aus dem fünfjährigen Krieg* (Berlin: Ullstein, 1920), pp. 412–13.

10. Germany, Nationalversammlung, *Stenographische Berichte*, II, 160. Also, Bethmann-Hollweg in a memorandum to Hindenburg pointed out that an American declaration of war as a result of the submarine campaign was almost a certainty, and that the effect of the campaign on England was "a matter of uncertain estimates." Yet he failed to recognize the linkage between England's response and America's entry into the war. (*Ibid.*, pp. 183–84.)

11. "Wird so verfahren, dann, aber auch nur dann, wird der Schrecken in die Schiffahrt, in das englische Volk und in die Neutralen fahren, der den Erfolg des uneingeschränkten U-Boot-Krieges verbürgt. Ich erwarte diesen Erfolg mit Sicherheit innerhalb eines Zeitraumes von längstens fünf Monaten. Der Erfolg wird genügen, um England zu einem brauchbaren Frieden geneigt zu machen." (*Ibid.*, p. 266.)

12. *Ibid.*, p. 228.

13. Earl of Oxford and Asquith, *Memories and Reflections, 1852–1927*, II (Boston: Little, Brown, 1928), 176.

14. Germany, Nationalversammlung, *Stenographische Berichte*, II, 229.

15. *Ibid.*, I, 358–359.

16. Vaïnö Tanner, *The Winter War* (Stanford, Calif.: Stanford University Press, 1957), pp. 152, 155, 169, 171.

17. The Soviet ambassador in London kept the British government informed of the Soviet conditions for peace with Finland, as if to assure the

British that Finland's independence would not be jeopardized. And in their negotiations with the Russians, the Finns deliberately used the Western offer for aid as a threat. (Carl Gustav Mannerheim, *Erinnerungen* [Zurich: Atlantis, 1952], pp. 410, 412.) This must have seemed ominous to the Soviet government, given the rumors that the British might make peace with Hitler. Thus, the Soviet ambassador in London had reported a few months earlier that the British government might be willing to make peace with Germany under the right conditions. (See *Documents on German Foreign Policy, 1918–1945*, Series D, Vol. VIII [U.S. Department of State Publication, Washington, D.C., 1954], No. 285, p. 325.)

How fragile and ill-prepared this French-British threat was, however, is further illustrated by the fact that only a week prior to the climax of the Soviet-Finnish peace negotiations, Sweden and Norway had refused a French-British request that passage and cooperation be extended to their expeditionary force, even though the Allies offered "extensive military assistance" as protection against possible German retaliation. Johan Jörgen Holst, "Surprise, Signals, and Reaction," *Cooperation and Conflict: Nordic Studies in International Politics*, II (1) (1966), 38.

18. Hitler gave this warning to Molotov when the latter visited him November 13, 1940. See *Documents on German Foreign Policy, 1918–1945*, Series D, Vol. XI (U.S. Department of State Publication, Washington, D.C., 1960), pp. 555–57.

19. Tanner, *The Winter War*, pp. 222–35.

20. Andreas Hillgruber, *Hitler's Strategie: Politik und Kriegführung, 1940–1941* (Frankfurt: Bernard & Graefe, 1965), p. 548.

Support for the thesis that Hitler attacked Russia primarily in order to force England to make peace is to be found, in particular, in Hillgruber's very thorough study of this period (*ibid.*, pp. 224–25, 573–75). For some references to Hitler's attempts to make peace with England, from his victory over Poland until 1941, see L. Woodward, *British Foreign Policy in the Second World War* (London: HMSO, 1962), and Hillgruber, *Hitler's Strategie*, pp. 114–15.

Regarding Soviet economic aid to Germany prior to Hitler's attack, see Raymond James Sontag and James Stuart Beddie, eds., *Nazi-Soviet Relations, 1939–1941* (U.S. Department of State Publication 3023, Washington, D.C., 1948). The Nazi-Soviet trade agreement of February, 1940, provided for Soviet deliveries of raw materials amounting to 500 million Reichsmarks. The following year a new agreement was concluded, and the Soviets delivered till the very date of the German attack (*ibid.*, pp. 131–41, 318–19, 339–41).

21. U.S. Senate, 82d Congress (1), Hearings Before the Committee on Armed Services and the Committee on Foreign Relations, *Military Situa-*

140
3. PEACE THROUGH ESCALATION?

tion in the Far East (Washington, D.C., 1951), Part 1, p. 19, and Part 2, p. 733.

22. F. W. Deakin, *The Brutal Friendship: Mussolini, Hitler and the Fall of Italian Fascism* (New York: Harper & Row, 1962), p. 217.

23. France, Assemblée Nationale, No. 2344 (1947) *Les Evénements survenus en France de 1933 à 1945*, Parties II–IV, Documents sur la période 1936–1945 (Paris, 1951), p. 350.

24. *Ibid.*, pp. 353–54.

25. "Elle permet de porter un coup très grave, sinon décisif, a l'organisation militaire et économique soviétique." (All this time France was at war with Germany, not Russia!) Gamelin's report is reproduced in a collection, apparently authentic, of French General Staff documents captured and published by the Germans: German Foreign Office, *Die Geheimakten des französischen Generalstabes* (Berlin, 1941), pp. 211–15. (The capture of these documents is mentioned in Hitler's discussion with the Italian ambassador on July 1, 1940, reported in *Documents on German Foreign Policy, 1918–1945*, Series D, Vol. X, No. 73 [U.S. Department of State Publication, Washington, D.C., 1954].)

26. L. M. Chassin, "Un plan grandiose: L'attaque des petroles du Caucase en 1940," *Forces Aériennes Françaises* (December, 1961), pp. 821–49. For the French-British plans for a landing in Norway see T. K. Derry, *The Campaign in Norway* ("[Brit.] History of the Second World War, U.K. Military Series" [London: HMSO, 1952]), especially pp. 13–14.

CHAPTER IV. THE STRUGGLE WITHIN:

PATRIOTS AGAINST "TRAITORS"

The Stauffenberg quotation is from Joachim Kramarz, *Claus Graf Stauffenberg* (Frankfurt: Bernard & Graefe, 1965), p. 201.

1. The French word *trahison,* from which the English "treason" derives, retains the broader meaning of "treason" *plus* "betrayal" (as does the German word *Verrat*). Hence, in these languages, it is more appropriate than in English to use the same word for the *treason* of helping the enemy behind the back of one's government, and the *betrayal* of the national interest through the instigation or prolongation of hostilities behind the back of one's own appropriate national authorities.

2. Erich Ludendorff, *Urkunden der Obersten Heeresleitung über ihre Tätigkeit 1916/18* (Berlin, 1920), pp. 375–76.

3. U.S. Department of State, *Foreign Relations of the United States, 1917,* Supplement 1, pp. 57–58, 63, 64, 65. Secret negotiations actually took place

between Austrian and Allied emissaries on several occasions. But Czernin and the other Austrian leaders always saw to it that there should be no "treason" against the German ally. See, for instance, G. de Manteyer, *Austria's Peace Offer, 1916–1917* (London: Constable, 1921).

4. Carnegie Endowment for International Peace, *Preliminary History of the Armistice* (New York: Oxford University Press, 1924), p. 29. Fritz Fischer, *Griff nach der Weltmacht: Die Kriegszielpolitik des Kaiserlichen Deutschland, 1914–1918* (Düsseldorf: Droste, 1961), p. 847.

5. Ottokar Czernin, *In the World War* (London: Cassell, 1919), pp. 217–18.

6. *Documents and Statements Relating to Peace Proposals and War Aims, December, 1916–November, 1918* (New York: Macmillan, 1919), pp. 210–11.

7. Carl Gustav Mannerheim, *Erinnerungen* (Zurich: Atlantis, 1952), pp. 495–96. Anatole C. Mazour, *Finland Between East and West* (Princeton, N.J.: Van Nostrand, 1956), pp. 160–68.

8. Mannerheim, *Erinnerungen*, p. 526. (Incidentally—lest some specialist be disturbed about my historical license—the Finns were never *de jure* allies of Nazi Germany.)

9. *Ibid.*, pp. 410, 414. The question deserves to be raised whether Mannerheim could have done more in 1941 to prevent Finland from joining Hitler's attack against Russia. This is not the place to offer a complete evaluation of this complicated story.

10. Edward Spears, *Assignment to Catastrophe*, Vol. I, *Prelude to Dunkirk, July 1939–May 1940* (New York: Wynn, 1954), p. 149.

11. German Foreign Office, *Die Geheimakten des französischen Generalstabes* (Berlin, 1941). (See chap. 3, note 25, regarding this source.)

12. Spears, *Assignment to Catastrophe*, II, 163.

13. *Ibid.*, pp. 148, 151.

14. Hermann Böhme, *Entstehung und Grundlagen des Waffenstillstandes von 1940* (Stuttgart: Deutsche Verlags-Anstalt, 1966), pp. 17, 20–22, 71–75.

15. Harry Rudin, *Armistice 1918* (New Haven: Yale University Press, 1944), pp. 246–51; and Hans Kutscher, *Admirals Rebellion oder Matrosenrevolte?* (Stuttgart: Kohlhammer, 1933), especially pp. 58–61.

16. André Tardieu, *The Truth about the Treaty* (Indianapolis: Bobbs-Merrill, 1921), p. 62.

17. Toshikazu Kase, *Journey to the Missouri* (New Haven: Yale University Press, 1950). Lester Brooks, *Behind Japan's Surrender: The Secret Struggle That Ended an Empire* (New York: McGraw-Hill, 1968), pp. 303-50.

18. For a fictionalized account of how General von Choltitz surrendered Paris, thus saving it from major destruction, see Larry Collins and Dominique Lapierre, *Is Paris Burning?* (New York: Simon & Schuster, 1965).

Allen Dulles recorded his role in negotiating the 1945 surrender of the German forces in Italy in his book, *The Secret Surrender* (New York: Harper & Row, 1966).

For an account of the final weeks of Hitler's Germany see Cornelius Ryan, *The Last Battle* (New York: Simon & Schuster, 1966), with its extensive bibliography; H. R. Trevor-Roper, *The Last Days of Hitler* (New York: Berkeley Publishing Co., 1957); and Albert Speer, *Inside the Third Reich* (New York: Macmillan, 1970).

19. In the same spirit, the West German military code has a clause stating that if a soldier fails to carry out an order that "violates human dignity," he does not become guilty of insubordination. (*Soldatengesetz* [March 19, 1956] 11 [1].)

On the Twentieth of July Movement see: Gerhard Ritter, *The German Resistance: Carl Goerdeler's Struggle Against Tyranny* (New York: Praeger, 1958); *Spiegelbild einer Verschwörung: Die Kaltenbrunner Berichte an Bormann und Hitler über das Attentat vom 20. Juli 1944. Geheime Dokumente aus dem ehemaligen Reichssicherheitshauptamt* (Stuttgart: Seewald Verlag, 1961); Peter Hoffman, *Widerstand—Staatsstreich—Attentat: Der Kampf der Opposition gegen Hitler* (Munich: R. Piper, 1969).

20. An important study of German war aims in World War I is Fritz Fischer, *Griff nach der Weltmacht*. Fischer's study, which stirred considerable controversy in Germany, indicates that most of the civilian government leaders supported the annexationist war aims that the military favored. Cf. the review of Fischer's book by Klaus Epstein in *World Politics*, XV (October, 1962), 163-85.

21. John Wheeler-Bennett, *Brest-Litovsk: The Forgotten Peace* (London: Macmillan, 1956), p. 263.

22. John Buchan, *History of the Great War* (Boston: Houghton Mifflin, 1923), III, 414.

23. One of the few books criticizing England's "hawkish" attitude in World War I is Irene Willis, *England's Holy War* (New York: Knopf, 1928).

24. *Parliamentary Debates, House of Commons*, Series 5, Vol. 73, Cols. 988, 1492.

25. *Ibid.*, Series 5, Vol. 75, Cols. 1558-60, 1563.

26. *Ibid.*, Series 5, Vol. 85, Col. 136. For a general account of the inadequacy of peace negotiations during World War I, see Kent Forster, *The Failures of Peace: The Search for a Negotiated Peace During the First World War* (Washington, D.C.: American Council on Public Affairs, 1941). On the British Parliament, see especially pp. 18–19, 33.

27. Earl of Oxford and Asquith, *Memories and Reflections, 1852–1927* (Boston: Little, Brown, 1928), II, 166, 170–71.

28. Gerda D. Crosby, *Disarmament and Peace in British Policy, 1914–1919* (Cambridge, Mass.: Harvard University Press, 1957), pp. 53–54.

29. Forster, *Failures of Peace*, pp. 52–57.
   A similar reproach, although based on more speculative evidence, has been made against British and American policy in World War II. Some writers have argued that if the British government, in particular, had given greater encouragement to the German resistance against Hitler, the war might have been shortened. More generally, the British reluctance to support the abortive revolt against Hitler has been criticized on moral grounds. David Astor criticizes "the narrowness and inadequacy" of the British wartime attitude, as exemplified by the British government's "flat refusal in the summer of 1944 to promise even a temporary cessation of the bombing of Berlin (which could have had no military significance) when the German opposition asked for this minimal sign of approval to follow, if they managed to kill Hitler and break his government. There has been almost no British criticism of this attitude in the twenty-five years that have since passed." "The German Opposition to Hitler," a letter in *Encounter*, October, 1969. See also Christopher Sykes, "Heroes and Suspects: The German Resistance in Perspective," *Encounter*, December, 1968, and David Astor, "Why the Revolt Against Hitler Was Ignored," *Encounter*, June, 1969.

30. Lloyd George, *War Memoirs*, Vol. III (1915–16) (Boston: Little Brown, 1933), p. 314. Actually, Lloyd George was not the most extreme "hawk" in the British government. He was not a Clemenceau, despite his famous call for the "knockout blow." His more moderate role at the Paris Peace Conference is well known. Also, he favored a separate peace with Austria. And in December, 1916, when he succeeded Asquith as Prime Minister, he assumed a slightly more conciliatory stance toward peace negotiations with Germany even though he depended for political support on the militant group. See Ernest R. May, *The World War and American Isolation, 1914–1917* (Cambridge, Mass.: Harvard University Press, 1959), pp. 381–82.

31. Czernin, *In the World War*, p. 247.

32. Wheeler-Bennett, *Brest-Litovsk*, p. 278. The statement by Bukharin and Radek appeared in the first issue of *Kommunist*.

CHAPTER V. THE STRUGGLE WITHIN:
SEARCH FOR AN EXIT

The full text of Emperor Hirohito's statement to his War Cabinet at the Imperial Conference of August 14, 1945, is reproduced in Shimomura Kainan, *Shusen hishi*, in Gaimushō [Japanese Ministry of Foreign Affairs], *Shūsen Shiroku* (Tokyo, 1952).

1. Since World War II the "unconditional surrender" formula has been severely criticized by some writers. For a balanced discussion of this issue, see Paul Kecskemeti, *Strategic Surrender: The Politics of Victory and Defeat* (Stanford, Calif.: Stanford University Press, 1958), pp. 223–41. Kecskemeti's book discusses the French, Italian, German, and Japanese surrenders in World War II and offers a lucid analysis of the critical bargain involved in any surrender negotiations, particularly as seen from the point of view of a rational strategy for nations as unitary actors.

Implications for war termination of the fact that nations are not unitary actors are dealt with in Robert Rothstein, "Domestic Politics and Peacemaking: Reconciling Incompatible Imperatives," *Annals of the American Academy of Political and Social Science* (November, 1970), pp. 62–75; Robert Randle, "The Domestic Origins of Peace," *ibid.*, pp. 76–85; and Morton H. Halperin, "War Termination as a Problem in Civil-Military Relations," *ibid.*, pp. 86–95.

For a more extensive and systematic treatment of international negotiation than that offered in this chapter, see Fred Charles Iklé, *How Nations Negotiate* (New York: Harper & Row, 1964).

2. Henri Carré, *Les Grandes Heures du Général Pétain: 1917* (Paris, 1952), pp. 156–58.

3. Edgar O'Ballance, *The Greek Civil War, 1944–1949* (London: Faber & Faber, 1966), pp. 191, 201.

4. Walter G. Hermes, *Truce Tent and Fighting Front* ("United States Army in the Korean War" [U.S. Army, Office of Chief of Military History, Washington, D.C.: U.S. Government Printing Office, 1966]), p. 177.

5. In his memoirs Eisenhower wrote of the American options for ending the war in Korea: "One possibility was to let the Communist authorities understand that, in the absence of satisfactory progress, we intended to move decisively without inhibition in our use of weapons, and would no longer be responsible for confining hostilities to the Korean Peninsula. We would not be limited by any world-wide gentleman's agreement. In India and in the Formosa Straits area, and at the truce negotiations at Panmunjom, we dropped the word, discreetly, of our intention. We felt quite sure it would reach Soviet and Chinese Communist ears." Dwight D. Eisenhower, *Mandate for Change* (Garden City, N.Y.: Doubleday, 1963), p. 181.

6. Hermes, *Truce Tent and Fighting Front*, pp. 442–46, 454–55.

7. Kent Forster, *The Failures of Peace: The Search for a Negotiated Peace During the First World War* (Washington, D.C.: American Council on Public Affairs, 1941), pp. 104–5.

8. Lester Brooks, *Behind Japan's Surrender: The Secret Struggle That Ended an Empire* (New York: McGraw-Hill, 1968), pp. 214 ff.; U.S. Department of State, *Foreign Relations of the United States: The Conference of Berlin (Potsdam), 1945*, I, 897–99, and II, 1265–69 and *passim*. Cf. Admiral W. Leahy, *I Was There* (New York: McGraw-Hill, 1950), p. 434: "Some of those around the President wanted to demand [Emperor Hirohito's] execution."

9. Brooks, *Behind Japan's Surrender*, pp. 132–33, 138–59.

10. D. George Kousoulas, *Revolution and Defeat: The Story of the Greek Communist Party* (London: Oxford University Press, 1965), pp. 252–54, 274–76; O'Ballance, *Greek Civil War*, p. 185.

11. A. J. P. Taylor, *The Origins of the Second World War* (New York: Atheneum, 1962), p. 92.

12. Leon Wolff, *In Flanders Fields: The 1917 Campaign* (New York: Viking Press, 1958), pp. 259–60. Liddell Hart's *Through the Fog of War* (New York: Random House, 1938) offers a subtle comparison of the widely differing interpretations of the battles of World War I to be found among the memoirs and official histories on the Allied side alone.

The Asquith quotation is cited in Lord Beaverbrook, *Politicians and the War, 1914–1916* (London: T. Butterworth, 1928), p. 69.

13. Germany, Deutsche Nationalversammlung, 1919–20, Untersuchungsausschuss über die Weltkriegsverantwortlichkeit, *Beilagen zu den Stenographischen Berichten . . .* , 2. Untersuchungsausschuss, Beilage 1 (Berlin, 1920), II, 320.

14. Carnegie Endowment for International Peace, *Preliminary History of the Armistice* (New York: Oxford University Press, 1924), p. 40.

15. Fritz Fischer, *Griff nach der Weltmacht: Die Kriegszielpolitik des Kaiserlichen Deutschland, 1914–1918* (Düsseldorf: Droste, 1961), p. 853.

16. Carnegie Endowment for International Peace, *Preliminary History*, pp. 98–99.

17. *Ibid.*, p. 105.

18. Antole G. Mazour, *Finland Between East and West* (Princeton, N.J.: Van Nostrand, 1956), pp. 158, 160, 164, 167.

19. Edward Spears, *Assignment to Catastrophe*, Vol. II, *Fall of France, June 1940* (New York: Wynn, 1954), pp. 308–9, 315, 306.

20. Cf. Albert N. Garland and Howard McGraw Smyth, *Sicily and the Surrender of Italy* ("U.S. Army in World War II, The Mediterranean Theatre of Operations" [Washington, D.C.: U.S. Government Printing Office, 1965]).

21. F. W. Deakin, *The Brutal Friendship: Mussolini, Hitler and the Fall of Italian Fascism* (New York: Harper & Row, 1962), pp. 248–49.

22. *Ibid.*, pp. 374–75, 378.

23. *Ibid.*, p. 407.

EPILOGUE: ENDING WARS BEFORE THEY START

1. Robert F. Kennedy, who quoted this sentence from Khrushchev's letter of October 26, 1962, wrote: "There was no question that the letter had been written by him personally." *Thirteen Days: A Memoir of the Cuban Missile Crisis* (New York: W. W. Norton, 1969), pp. 86–87.

2. In December, 1942, Mussolini instructed Italian Foreign Minister Ciano to tell Hitler that he considered it "extremely advisable to come to an agreement with Russia." (F. W. Deakin, *The Brutal Friendship: Mussolini, Hitler and the Fall of Italian Fascism* [New York: Harper & Row, 1962], pp. 87, 94.)

3. Great Britain, *Documents on British Foreign Policy, 1919–1939* (London, 1949), Third Series, II (1938), 531–32.

The quotation from Nevile Henderson at the beginning of this section is from a telegram he sent to his Foreign Secretary, cited *ibid.*, IV (1939), 593.

4. *Ibid.*, pp. 526–27.

5. *Ibid.*, pp. 528, 269.

6. R.A.C. Parker, "The First Capitulation: France and the Rhineland Crisis of 1936," *World Politics*, VIII (1956), 355–73.

7. *Ibid.*, p. 364.

8. An "opération de coércition," according to the French planning documents. France, Assemblée Nationale, No. 2344, Aug. 8, 1947, *Les Evénements survenus en France de 1935 à 1945*, Part I by Charles Serre (Paris, 1951), pp. 30–38.

9. There seems to be no reliable evidence to support the contention that Hitler would have given in, if not fallen from power, as soon as France moved her little finger. (General Jodl's testimony at the Nuremberg trials [XV, 352] is dubious evidence, given its forensic purpose of minimizing aggressive intentions.) A massive military action by France, backed by Great Britain, would have been another matter. If one is willing to disregard the question whether French public and governmental opinion, let

alone the British government, was ready for such an overwhelming counter-move, one may judge Winston Churchill right in believing that "if the French Government had mobilized the French Army, with nearly a hundred divisions, and its air force . . . there is no doubt that Hitler would have been compelled by his own General Staff to withdraw, and a check would have been given to his pretensions which might well have proved fatal to his rule." (*The Gathering Storm* [Boston: Houghton Mifflin, 1948], p. 194.)

10. Great Britain, *Documents on British Foreign Policy, 1919–1939*, Third Series, IV (1939), 595. (This telegram was sent on March 15, 1939.)

11. Andreas Hillgruber, *Hitler's Strategie: Politik und Kriegführung 1940–1941* (Frankfurt: Bernard & Graefe, 1965), p. 40, n. 57.

12. Great Britain, *Documents on British Foreign Policy, 1919–1939*, Third Series, V (1939), 478.

13. Winston Churchill delivered this speech in the House of Commons on March 1, 1955. The speech also contains the warning about "dictators in the mood of Hitler" (cited at the beginning of this section), to which Churchill added, however, that "happily, we may find methods of protecting ourselves, if we were all agreed, against that," without explaining, though, what remedies he had in mind.

The quotation from Adolf Hitler is from a conversation during Hitler's first years in power, as recalled by Hermann Rauschning, *Gespräche Mit Hitler* (New York: Europa Verlag, 1940), p. 11.

Excellent overviews of the evolution of nuclear strategy since World War II are offered by Michael Howard, "The Classical Strategists," *Adelphi Papers* 54 (London: Institute for Strategic Studies, 1969), pp. 18–32, and Raymond Aron, "The Evolution of Modern Strategic Thought," *ibid.*, pp. 1–17.

14. President Nixon's message, *United States Foreign Policy for the 1970's* (February, 1970).

15. See, for instance, Herman Kahn, *On Thermonuclear War* (Princeton, N.J.: Princeton University Press, 1960), pp. 185, 302; and Bernard Brodie, *Strategy in the Missile Age* (Princeton, N.J.: Princeton University Press, 1959), pp. 293–94.

16. During the 1960s, the possibility that a surprise attack might sufficiently disarm its victim to render the damage from retaliation "acceptable" to the aggressor became a central concern of American nuclear strategy and seems to be the cardinal consideration in the American approach to strategic arms control. Whether or not this possibility is adequately foreclosed, however, depends in a complicated fashion both on various vulnerabilities of strategic forces and on the aggressor's capability to defend himself against a retaliatory strike. For a lucid explanation of how deterrence

against a disarming attack requires careful design of the retaliatory forces, see Albert Wohlstetter, "The Delicate Balance of Terror," *Foreign Affairs*, XXXVII (1959), 211–34. The possibility that the incentive to preempt the adversary's surprise attack might lead to a failure of deterrence is dealt with in Thomas C. Schelling's "The Reciprocal Fear of Surprise Attack," in his *The Strategy of Conflict* (Cambridge, Mass.: Harvard University Press, 1960).

17. Senator J. W. Fulbright on March 26, 1969, in *Hearings: Strategic and Foreign Policy Implications of ABM Systems*, U.S. Senate, Subcommittee on International Organization and Disarmament Affairs of the Committee on Foreign Relations, 1969, 314–15.

In his telling critique of this and related proposals, Paul Wolfowitz reminds us that "it would be unconscionable to neglect to provide every possible means of limiting even nuclear hostilities, and of preventing—if only at the last moment—the total destruction of two great nations." (Paul W. Wolfowitz, "The Proposal to Launch on Warning," *Hearings: Military Procurement for Fiscal Year 1971*, U.S. Senate, Committee on Armed Services, 1970, pp. 2278–81.)

18. Senators Albert Gore and Stuart Symington, and Dr. Jerome Wiesner, among others. See Paul Wolfowitz, *Hearings: Military Procurement for Fiscal Year 1971*, p. 2280.

19. For example, a Soviet military writer, Lieutenant Colonel E. Rybkin, wrote in 1965: "To assert that victory is not at all possible in a nuclear war would not only be untrue on theoretical grounds but dangerous as well from the political point of view. . . . [One] can achieve a quick victory over the aggressor that will prevent further destruction and calamities. There are opportunities to create and develop new means of conducting war that are capable of reliably countering an enemy's nuclear blows." Cited in Roman Kolkowitz, *The Red "Hawks" on the Rationality of Nuclear War* (Santa Monica, Calif.: The Rand Corporation, 1966), p. 46.

Soviet public discussions of the doctrine of preemption tend to be opaque and understandably guarded. But the notion keeps recurring in Soviet military writings. See Herbert S. Dinerstein, Leon Gouré, and Thomas W. Wolfe, Introduction to V. D. Sokolovski, *Soviet Military Strategy* (Englewood Cliffs, N.J.: Prentice-Hall, 1963), pp. 63–66.

20. For a comprehensive presentation of publicly available data on past incidents and accidents affecting nuclear weapons and on the development of safeguards for U.S. weapons systems, see Joel Larus, *Nuclear Weapons Safety and the Common Defense* (Columbus, Ohio: Ohio State University Press, 1967), especially chap. 2.

21. W. Warlimont, *Im Hauptquartier der deutschen Wehrmacht, 1939–1945* (Frankfurt: Bernard & Graefe, 1962), pp. 221–22.

22. D. G. Brennan, "The Case for Missile Defense," *Foreign Affairs*, XLVII (1969), 433–48, offers an analysis of how American views on active missile defenses evolved in opposition to the initial Soviet views. This article develops a well-reasoned critique of the sole reliance, for purposes of deterrence, on "assured destruction"—the strategy that guarantees an *unlimited* "assured vulnerability" of cities to any form of nuclear attack.

BIBLIOGRAPHY ON THE TERMINATION
OF WARS

As mentioned in the Preface, the literature on how wars are brought to an end is much smaller than the literature on the initiation of wars or on the conduct of military campaigns. The tendency of historians, political scientists, and strategists to neglect the question of war termination is discussed by William T. R. Fox, "The Causes of Peace and Conditions of War," *Annals of the American Academy of Political and Social Science* (November, 1970), pp. 2–5; and Berenice A. Carroll, "War Termination and Conflict Theory: Value Premises, Theories, and Policies," *ibid.*, pp. 14–29. The latter article contains references to recent American writings on war termination. For further references and an elucidation of taxonomic questions, see Berenice A. Carroll, "How Wars End: An Analysis of Some Current Hypotheses," listed below.

Abt, Clark C. "The Termination of General War." Ph.D. thesis, M.I.T., Cambridge, Mass., January, 1965.

Aron, Raymond. *Paix et Guerre.* Paris: Calmann-Levy, 1962.

Böhme, Hermann. *Entstehung und Grundlagen des Waffenstillstandes von 1940.* Stuttgart: Deutsche Verlags-Anstalt, 1966.

Brooks, Lester. *Behind Japan's Surrender: The Secret Struggle That Ended an Empire.* New York: McGraw-Hill Book Co., 1968.

Butow, R. J. C. *Japan's Decision to Surrender.* Stanford, Calif.: Stanford University Press, 1954.

Calahan, Lt. Cmdr. H. A. (USNR). *What Makes a War End?* New York: Vanguard Press, Inc., 1944.

Carroll, Berenice A. "How Wars End: An Analysis of Some Current Hypotheses," *Journal of Peace Research,* 4, 1969. International Peace Research Institute, Oslo.

Dahlin, Ebba. *French and German Public Opinion on Declared War Aims, 1914–1918*. Stanford University Publications, University Series, History, Economics, and Political Science, Vol. IV, No. 2. Stanford, Calif.: Stanford University Press, 1933.

Documents and Statements Relating to Peace Proposals and War Aims (December, 1916–November, 1918). New York: The Macmillan Co., 1919.

Dulles, Allen. *The Secret Surrender.* New York: Harper & Row, 1966.

Forster, Kent. *The Failures of Peace: The Search for a Negotiated Peace During the First World War.* Washington, D.C.: American Council on Public Affairs, 1941.

Garland, Albert N., and Howard McGraw Smyth. *Sicily and the Surrender of Italy.* U.S. Army in World War II, The Mediterranean Theatre of Operations. Washington, D.C.: U.S. Government Printing Office, 1965.

Gilbert, Martin, and Richard Gott. *The Appeasers.* Boston: Houghton Mifflin Co., 1963.

Grew, J. C. *Turbulent Era: A Diplomatic Record of Forty Years, 1904–1945.* Ed. by Walter Johnson. 2 vols. Boston: Houghton Mifflin Co., 1952.

Halperin, Morton. *Limited War in the Nuclear Age.* New York: John Wiley, 1963.

Hentig, Hans von. *Der Friedensschluss: Geist und Technik einer Verlorenen Kunst.* Stuttgart: Deutsche Verlags-Anstalt, 1952.

Hermes, Walter G. *Truce Tent and Fighting Front.* United States Army in the Korean War. U.S. Army, Office of Chief of Military History, Washington. D.C.: U.S. Government Printing Office, 1966.

Hölzle, Erwin. "Das Experiment des Friedens im Ersten Weltkrieg, 1914–1917," *Geschichte in Wissenschaft und Unterricht,* XIII (August, 1962), 465–522.

"How Wars Are Ended," *Annals of the American Academy of Political and Social Science,* November, 1970. Ed. by William T. R. Fox.

Kahn, Herman. *On Escalation: Metaphors and Scenarios.* New York: Praeger, 1965.

Kahn, Herman, William Pfaff, and Edmund Stillman. "War Termination Issues and Concepts." Harmon-on-Hudson, N.Y.: Hudson Institute, June, 1968.

Kase, Toshikazu. *Journey to the Missouri.* New Haven: Yale University Press, 1950.

Kecskemeti, Paul. *Strategic Surrender: The Politics of Victory and Defeat.* Stanford, Calif.: Stanford University Press, 1958.

Klingberg, F. L. "Predicting the Termination of War: Battle Casualties and Population Losses," *Journal of Conflict Resolution,* June, 1966, pp. 129–71.

Knorr, Klaus. *On the Uses of Military Power in the Nuclear Age.* Princeton, N.J.: Princeton University Press, 1966.

Lanyi, G.A. "The Problem of Appeasement," *World Politics*, XV (January, 1963), 316–28.

Manteyer, G. de. *Austria's Peace Offer, 1916–1917*. London: Constable and Company, Ltd., 1921.

May, E. R. *The World War and American Isolation, 1914–1917*. Cambridge, Mass.: Harvard University Press, 1959.

Mayer, A. T. *Political Origins of the New Diplomacy, 1917–1918*. New Haven: Yale University Press, 1959.

—— *Politics and Diplomacy of Peacemaking: Containment and Counterrevolution at Versailles, 1918–1919*. New York: Alfred A. Knopf, Inc., 1967.

Mazour, Anatole C. *Finland Between East and West*. Princeton, N.J.: D. Van Nostrand Company, Inc., 1956.

Mourin, Maxime. *Les Tentatives de Paix dans la Seconde Guerre Mondiale (1939–45)*. Paris: Payot, 1949.

Mowat, R. B. *Diplomacy and Peace*. London: Williams & Norgate, 1935.

Osgood, R. E. *Limited War: The Challenge to American Strategy*. Chicago: University of Chicago Press, 1957.

Phillipson, Coleman. *Termination of War and Treaties of Peace*. New York: E. P. Dutton & Co., 1916.

Pitt, Barrie. *1918: The Last Act*. New York: W. W. Norton and Company, Inc., 1962.

Renouvin, Pierre. *L'Armistice de Rethondes: 11 Novembre 1918*. Paris: Gallimard, 1968.

Rudin, Harry R. *Armistice 1918*. New Haven: Yale University Press, 1944.

Schelling, Thomas C. *Arms and Influence*. New Haven: Yale University Press, 1966.

Scherer, André, and Jacques Grunewald. *L'Allemagne et les problemes de la paix pendant la premiere guerre mondiale*. Vol. I: August, 1914–January, 1917. Paris: Presses Universitaires, 1962.

Shūsen Shiroku [History of the Termination of the War]. Compiled by Gaimushō [Japanese Ministry of Foreign Affairs]. Tokyo, 1952.

Tardieu, André, *The Truth about the Treaty*. Indianapolis: Bobbs-Merrill Company, 1921.

Tompkins, Peter. *Italy Betrayed*. New York: Simon and Schuster, Inc., 1966.

Trever-Roper, H. R. *The Last Days of Hitler*. New York: Berkeley Publishing Corp., 1957.

Upton, Anthony F. *Finland in Crisis, 1940–1941: A Study in Small Power Politics*. Ithaca, N.Y.: Cornell University Press, 1964.

Willis, Irene C. *England's Holy War: A Study of English Liberal Idealism During the Great War*. New York: Alfred A. Knopf, Inc., 1928.

INDEX

Abyssinia: Italian invasion, 97
Algeria, 24; and France, war, 12, 22, 68-69, 87
Allison, Graham T., 134n
Alsace, 21-22
Ambrosio, Vittorio, 103-4
Appeasement and conciliation, 7-8, 108, 110, 112, 114-16, 118
Arabs: and Israel, 19, 25, 109; see also Egypt
Arms control, 108, 121, 128, 129
Aron, Raymond, 146n
Asquith, Herbert Henry (Earl of Oxford and Asquith), 76-77, 78, 79, 97
Astor, David, 142n
Australia: World War II, 93
Austria: World War I, 9, 21, 32, 49, 62-64, 65, 75, 82, 92-93, 109-10; and Germany, 10; and Serbia, 40

Badoglio, Pietro, 103
Balfour, Arthur James (Earl of Balfour), 11
Beaufre, André, 7
Belgium: World War I, 42
Berlin, battle of, 37
Bethmann-Hollweg, Moritz August von, 8-9, 99, 100, 101

Birnbaum, Karl E., 137n
Bismarck, Otto von, 10, 11
Bohemia, 63
Böhme, Hermann, 141n
Bombing, effect of, 28, 29-30
Bradley, Omar, 54
Brennan, Donald G., 148n
Brodie, Bernard, 147n
Brooks, Lester, 141n, 144n
Bukharin, Nikolai, 83
Bulgaria: World War I, 32

Canaris, Wilhelm, 34
Cecil, Lord Robert, 48-49
Chamberlain, Neville, 9-10, 110-11
China: and Japan, war, 2, 3
China, Communist, 130; and Tibet, 6; Korean war, 22-23, 27, 32, 36, 37, 54-55, 88-92; nuclear weapons, 120, 121; and Soviet Union, border clashes, 124
Choltitz, Dietrich von, 72
Chou En-lai, 88
Churchill, Winston, 19, 29, 33, 67, 93, 118, 119, 146n
Ciano, Conte Galeazzo, 145n
Civil war, 95
Clark, Mark, 92
Clausewitz, Karl von, 17

Concessions: in preventing war, *see* Appeasement and conciliation; in terminating war, 59-60, 66, 69, 74, 83, 85

Conciliation, *see* Appeasement and conciliation

Crosby, Gerda D., 142*n*

Cuban missile crisis, 106

Czechoslovakia: German invasion, 110, 111, 112, 115

Czernin, Count Ottokar, 62-63, 64, 82

Daladier, Edouard, 51, 57, 110, 111-12

Darlan, Jean Louis, 102

Deakin, F. W., 136*n*, 139*n*, 145*n*

Denmark: World War I, 44

Deterrence, 108, 114-15, 118-30

Dewey, John, vi

Disarmament, 108, 121, 128

Dulles, Allen, 72, 94

Eden, Anthony: Suez crisis, 6-7, 114, 116-17

Egypt: Suez crisis, 6-7, 114, 116-17; Six-Day War, 19

Eisenhower, Dwight D., 91, 92

Epstein, Klaus W., 133*n*, 141*n*

Ethiopia, *see* Abyssinia

Falkenhayn, Erich von, 8

Finland: Finnish-Soviet war (1940; Winter War), 5-6, 25, 41, 50-52, 57, 66, 86; World War II, 25-26, 64-66, 86, 101-2

Fischer, Fritz, 133*n*, 136*n*, 140*n*, 141*n*, 145*n*

Foch, Ferdinand, 71

Forrestal, James, 136*n*

Forster, Kent, 142*n*, 144*n*

France: Suez crisis, 6-7; World War I, 8, 9, 21, 31, 33, 42, 45, 46, 63, 87, 92-93, 100, 130; Franco-Prus-

sian War, 10-11; World War II, 11, 18-19, 23-24, 41, 53, 57-58, 66-68, 102, 111; and Algeria, war, 12, 22, 68-69, 87; Finnish-Soviet war (1940), 50, 51, 57; appeasement (and Munich Conference), 110-15 *passim*, 117; nuclear deterrence, 120

Franco, Francisco, 24

Franco-Prussian War, 10-11

Fulbright, J. W., 147*n*

Gamelin, Maurice Gustave, 58, 113

Garland, Albert N., 135*n*, 145*n*

Gaulle, Charles de, 12, 68, 87

Genocide, 130-31

Germany, 11; World War I, 8-9, 10, 20-21, 28-29, 31-32, 33, 36, 37, 42-49, 53, 58, 62-64, 70-71, 73, 75, 82-83, 86-87, 92, 98-101, 109-10, 130, 133*n*-134*n*; World War II, 9-10, 19, 20-21, 23-24, 25, 29-30, 33, 34, 37, 41, 50-51, 52-53, 56, 57, 65, 66-68, 71-74, 77, 80-81, 82, 86, 87, 101, 102, 103, 104, 107, 117-18, 127-28, 130; appeasement (and Munich Conference), 9-10, 110-15 *passim*, 117; and Austria, 10; Franco-Prussian War, 10-11; Nazi-Soviet Pact, 50, 51, 58, 107, 117, 139*n*; and Saar, 109, 113; Czechoslovakia invaded, 110, 111, 112, 115; Rhineland, 113, 114; *see also* Hitler, Adolf

Gibraltar: World War II, 24, 56

Great Britain: Suez crisis, 6-7, 114, 116-17; World War I, 9, 11, 21, 28-29, 31-32, 33, 42-49 *passim*, 63, 70-71, 75-82, 87, 92, 97, 99; appeasement (and Munich Conference), 9, 110-15 *passim*, 117; World War II, 9-10, 11, 19, 24, 25, 26, 29-30, 34, 36, 52-53, 56, 57-58, 66, 67, 81, 82, 97, 117-18,

142n; U.S. aid, 25; Korean war, 36; Finnish-Soviet war (1940), 50, 51, 57; and Italian invasion of Abyssinia, 97; nuclear deterrence, 119, 120
Greece: World War I, 43; civil war, 88, 95

Haig, Douglas, 76, 97
Halifax, Lord, 115
Halperin, Morton H., 134n, 136n, 143n
Hart, Liddell, 144n
Henderson, Sir Nevile, 109, 115
Hermes, Walter G., 134n, 143n, 144n
Hillgruber, Andreas, 138n, 146n
Himmler, Heinrich, 72
Hindenburg, Paul von, 43, 99, 100, 101, 137n
Hirohito, Emperor, 2, 71, 84, 93, 103
Hiroshima, 55, 118
Hitler, Adolf, 9, 10, 21, 23, 24, 25, 26, 37, 50-51, 52, 56, 59, 65, 66, 67, 68, 71, 72-74, 82, 87, 102, 103, 104, 107-8, 110-15 passim, 117, 118, 127, 130, 139n, 142n
Hoffman, Peter, 141n
Holst, Johan Jörgen, 138n
House, Edward M., 75
Howard, Michael, 146n
Hungary: revolution (1956), 6; see also Austria: World War I
Hussein, King: Six-Day War, 19

Ike, Nobutaka, 133n
Iklé, Fred Charles, 135n, 143n
India: and Pakistan, war, 86
Israel, 109; Suez crisis, 6, 7; Six-Day War, 19; and Arabs, 25
Italy: World War I, 21, 92-93; World War II, 24, 25, 33, 34-35, 36, 56, 68, 86, 87, 92, 97, 103-5; Abys-

sinia invaded, 97; see also Mussolini, Benito

Japan: and China, war, 2, 3; World War II, 2, 3-4, 19, 21, 33-34, 55, 71, 73, 84, 86, 93-95, 99, 103; World War I, 33
Jodl, Alfred, 127-28, 146n
Johnson, Lyndon B., 38
Jordan: Six-Day War, 19; see also Arabs

Kahn, David, 135n
Kahn, Herman, 136n, 146n, 148n, 149n
Karl, Emperor, 62
Kase, Toshikazu, 141n
Kecskemeti, Paul, 143n
Keitel, Wilhelm, 117
Kennedy, John F., 106
Kennedy, Robert F., 145n
Keynes, J. M., 76
Khrushchev, Nikita S., 106
Klein, Burton H., 134n
Kolkowitz, Roman, 147n
Korea, 36, 37, 54, 55, 89, 95; see also North Korea; South Korea
Korean war, 22-23, 25, 27, 32, 36, 37, 54-55, 88-92, 95, 113n
Kousoulas, D. George, 144n
Kutscher, Hans, 141n

Lansdowne, Lord, 64, 79
Lansing, Robert, 63
Larus, Joel, 148n
Law, Bonar, 78
Leahy, William, 144n
Lebrun, Albert, 67
Lenin, V. I., 75, 82-83
Libya, 24, 109
Lincoln, Abraham, 1
Lindemann, Frederick, 29-30
Lloyd George, David, 48, 78, 80-81, 133n

Lorraine, 11, 21
Ludendorff, Erich, 31, 36, 44, 45, 82, 86-87, 99, 100

MacArthur, Douglas, 1, 38, 54, 89, 133n
Mannerheim, Carl Gustav, 26, 65-66, 102, 140n
Manteyer, G. de, 140n
May, Ernest R., 143n
Mazour, Anatole C., 134n, 140n, 145n
Mexico: World War I, 33
Mitterand, François, 12
Molotov, V. N., 138n
Morgenstern, Oscar, 136n
Morocco, 56
Munich Conference, 110-12, 115, 117
Mussolini, Benito, 25, 34-35, 36, 56, 68, 87, 97, 103-5, 107, 127

Nagasaki, 55, 118
Napoleon, 1
Nasser, Gamal Abdel: Suez crisis, 6, 7, 116; Six-Day War, 19
NATO, 120
Netherlands, 44
Neustadt, Richard E., 134n
Nixon, Richard M., 122-23
Non-proliferation Treaty, 120
North Africa: World War II, 23-24, 34, 56, 68
North Korea, 25, 89; see also Korea; Korean war
Norway: Finnish-Soviet war (1940), 41, 51, 57; World War II, 139n
Nuclear war, 27, 28, 40, 55, 91, 94, 107-8, 118-31 passim; deterrence, 119-31 passim; retaliation, 122, 125

O'Ballance, Edgar. 143n, 144n

Pakistan: and India, war, 86
Parker, R. A. C., 145n

Pearl Harbor, attack on, 2-3, 4, 5, 19, 33, 99
Pétain, Henri, 23, 66-68, 102
Pfaff, William, 148n
Poland, 11, 117; World War II, 10, 11, 50
Potsdam Conference, 33, 34, 93, 103

Radek, Karl, 83
Randle, Robert, 143n
Rationality, 14-16, 127
Reynaud, Paul, 57, 66, 67, 102
Rhee, Syngman, 37, 91-92
Rhineland, 113, 114
Ridgway, Matthew, 89-90
Ritter, Gerhard, 141n
Roosevelt, Franklin D., 33, 93
Rothstein, Robert, 143n
Rudin, Harry, 141n
Rumania, 117
Russia: World War I, 8, 21, 63, 64, 75, 82-83, 106, 109-10; see also Soviet Union
Ryan, Cornelius, 141n
Ryti, Rysto, 52, 65

Saar, 109, 113
Sallagar, F. M., 135n
Scheer, Reinhard, 70
Scheidemann, Philipp, 75
Schelling, Thomas C., 147n
Serbia: and Austria, 40
Smith, Gordon, 135n
Smyth, Howard McGraw, 135n, 145n
Snow, C. P., 29
South Korea, 25, 55, 89; see also Korea; Korean war
Soviet Union, 11, 27, 111, 128, 130; Finnish-Soviet war (1940; Winter War), 5-6, 25, 41, 50-52, 57, 66, 86; and Hungarian revolution (1956), 6; World War II, 24, 25-26, 33, 34, 52, 53, 55, 56, 57-58, 64-66, 86, 94-95, 101-2, 103, 106-

7, 117-18, 130; Korean war, 27, 88-92; Nazi-Soviet Pact, 50, 51, 58, 107, 117, 139n; and Yugoslavia, 88, 95; Cuban missile crisis, 106; nuclear deterrence, 119, 120, 124, 126; and Communist China, border clashes, 124; see also Russia; Stalin, Josef
Spain: World War II, 24, 56
Spears, Edward, 140n, 145n
Speer, Albert, 72
Stalin, Josef, 26, 33, 50, 51, 52, 53, 88, 89, 91, 93, 94-95, 106-7, 117
Stauffenberg, Claus von, 59
Stillman, Edmund O., 148n, 149n
Stimson, Henry, 34
Submarine warfare, 28-29, 42-49 passim, 58, 70, 80, 99
Suez crisis, 6-7, 114, 116-17
Sugiyama, Hajime, 2-3
Sweden: Finnish-Soviet war (1940), 50, 51; World War II, 101

Tardieu, André, 141n
Taylor, A. J. P., 96-97
Tibet: and Communist China, 6
Tito, Josip Broz, 88, 95
Tizard, Sir Henry, 29
Togo, Shigenori, 4-5
Trevelyan, Charles, 77-78, 82
Trevor-Roper, H. R., 141n
Trotsky, Leon, 75
Truman, Harry S., 33-34, 89, 91
Tsukada (Army Vice Chief of Staff), 4-5
Tuchman, Barbara, 135n
Tunisia, 24

Union of Soviet Socialist Republics, see Soviet Union
United Nations: Korean war, 22, 23, 32, 36, 37, 54, 55, 88-92 passim
United States, 128, 130; World War II, 2, 3-4, 19, 21, 26, 33-34, 36, 55, 56, 72, 84, 86, 93-95, 99, 103, 117-18, 127-28, 142n; Indian wars, 6; World War I, 11, 33, 43-49 passim, 58, 63, 75, 99; Korean war, 22-23, 27, 32, 36, 37, 54-55, 88-92, 133n; aid to Great Britain, 25; Cuban missile crisis, 106; nuclear deterrence, 119, 120-21, 124, 126
Upton, Anthony F., 134n

War: aims and reasons for, 1-2, 5, 6, 7, 10-11, 14-15, 49, 109, 124; military and military estimates, 1-2, 6, 13, 14, 15, 16, 18-20, 32-34, 35-36, 38, 40, 84, 109, 134n; costs and risks, 2, 7-8, 11-12, 20-32, 36, 38-39, 40, 42, 74, 81, 83, 95-96, 97, 98; outside intervention, 2, 6, 20, 23-27, 39, 40; timing and means vs. ends, 4, 6; problems in backing out of, 9-12; prolongation of, 11-12, 39, 40-41, 42, 61-62, 123-24; temporary lulls or cease-fires, 12, 87-88; resources and mobilization, 16, 18, 20-23, 39, 40; nuclear war, 27, 28, 40, 55, 91, 94, 107-8, 118-31; effect of bombing, 28, 29-30; submarine warfare, 28-29, 42-49 passim, 58, 70, 80, 99; geographic expansion, 39, 40, 41; efforts to limit, 39-42, 52, 55-56; escalation, 39-42, 52, 55-56; self-justification for failure, 49-50; "war crimes," 61; civil war, 95; genocide, 130-31
—— prevention: appeasement and conciliation, 7-8, 108, 110, 112, 114-16, 118; as aim, 9-10, 106; elimination of sources of conflict, 108; outcome worse than price of peace, 108; deterrence, 108, 114-15, 118-30; disarmament, 108, 121, 128; arms control, 108, 121,

War: prevention (*continued*) 128, 129; changes in views and strategy, 128-29

—— termination: bring greater security than before, 9; negotiations while fighting continues, 12, 85-86, 87-88; effect on government and internal politics, 13, 59-61, 66, 67-68, 69, 74, 83, 84, 86, 97-98, 102, 134n; military prospects reassessed, 13, 95, 96; problems, 15-16; how to force enemy to surrender or bargain, 17-18; fighting until bargaining power lost, 34; differing opinions, 35; aims reassessed, 36, 49, 74, 81, 83, 95-96, 97-98; depending on outcome of a single battle, 37; choices indeterminate, 39; possible peace terms often obscure, 39; effect of damage on enemy's peace terms, 41-42; after army is defeated, 52; concessions, 59-60, 66, 69, 74, 83, 85; "treason" vs. patriotism, 60-62, 66, 73, 81-82, 83, 94; when victory seems out of reach, 85; communication with enemy government as obstacle, 94; effect of unsuccessful or stalemated war, 102

Wheeler-Bennett, John, 141n, 143n
Whiting, Allen, 135n
Wilhelm, Kaiser, 31-32, 43, 47, 70
Willis, Irene, 142n
Wilson, Woodrow, 11, 33, 70, 75, 79, 99, 100
Wohlstetter, Albert, 135n, 147n, 148n
Wohlstetter, Roberta, 135n
Wolff, Karl, 72
Wolff, Leon, 97
Wolfowitz, Paul, 147n
World War I, 8-9, 10, 11, 20-21, 28-29, 31-32, 33, 36, 37, 42-49, 53, 58, 62-64, 65, 70-71, 73, 74-83, 86-87, 92-93, 97, 98-101, 106, 109-10, 125, 129, 130, 133n-134n
World War II, 2, 3-4, 9-10, 11, 18-19, 20, 21, 23-24, 25-26, 28, 29-30, 33-37 *passim*, 41, 50-51, 52-53, 55, 56, 57-58, 60, 64-68, 71-74, 76, 81, 82, 84, 86, 87, 92, 93-95, 97, 99, 101-7 *passim*, 111, 117-18, 127-28, 130, 142n; unconditional surrender, 86, 93-94

Yugoslavia: and Soviet Russia, 88, 95

Zachariades, Nicholas, 95